T0135093

Intelligent Systems Reference Library

Volume 89

Series editors

Janusz Kacprzyk, Polish Academy of Sciences, Warsaw, Poland
e-mail: kacprzyk@ibspan.waw.pl

Lakhmi C. Jain, University of Canberra, Canberra, Australia and
University of South Australia, Adelaide, Australia
e-mail: Lakhmi.Jain@unisa.edu.au

About this Series

The aim of this series is to publish a Reference Library, including novel advances and developments in all aspects of Intelligent Systems in an easily accessible and well structured form. The series includes reference works, handbooks, compendia, textbooks, well-structured monographs, dictionaries, and encyclopedias. It contains well integrated knowledge and current information in the field of Intelligent Systems. The series covers the theory, applications, and design methods of Intelligent Systems. Virtually all disciplines such as engineering, computer science, avionics, business, e-commerce, environment, healthcare, physics and life science are included.

More information about this series at http://www.springer.com/series/8578

George Mengov

Decision Science:
A Human-Oriented
Perspective

 Springer

George Mengov
Faculty of Economics and Business Administration
Sofia University St. Kliment Òhridski
Sofia
Bulgaria

Parts of this book have previously been published in the Bulgarian language in the book "Decision Making under Risk and Uncertainty", published by Janet 45 Publishing, Plovdiv.

ISSN 1868-4394   ISSN 1868-4408   (electronic)
Intelligent Systems Reference Library
ISBN 978-3-662-50952-4   ISBN 978-3-662-47122-7   (eBook)
DOI 10.1007/978-3-662-47122-7

Springer Heidelberg New York Dordrecht London
© Springer-Verlag Berlin Heidelberg 2015
Softcover reprint of the hardcover 1st edition 2015

Printed on acid-free paper

Springer-Verlag GmbH Berlin Heidelberg is part of Springer Science+Business Media
(www.springer.com)

For my students from Sofia University
St. Kliment Òhridski
For my friends

Preface

How people take decisions is an enticing enigma. Humanity has sought to crack it for millennia but apart from the occasional insight by a sage, little was achieved until the enterprise of science became strong enough to take up the issue. With the invention of probability theory in the seventeenth century and the concept of subjective utility in the eighteenth, decision analysis was put on a pragmatic basis. Yet, its success ever since has been very modest compared with the accomplishments of the natural sciences. The main reason is of course the difficulty to conduct research on such an elusive subject. Decision—understood as making up one's mind about a new belief or about an act to be carried out—is indeed tricky to study.

At any moment in history, scientific advancement in this field depended crucially on the available methods to analyse human mind and behaviour. Initially, scholars had to resort only to observation and philosophising. Armed with calculus and probability, the Renaissance mathematicians were taken by the idea to examine the behaviour of gamblers. Soon, however, their attention shifted to the seafaring risks and insurance calculations. Thus, decision analysis developed a strong affinity for economics which lasts to this day. In the meantime, psychophysics—later subsumed by psychology—made a powerful entry into the scene, bringing in the experimental method with empirical data gathering. As all of these streaks seemed to be converging into a loosely coherent framework around utility theory, twentieth century psychology began discovering instances of human irrationality at a rate, by far outclassing the decision theorists' struggle to accommodate them. There is yet no sign that this state of affairs will change soon. Finally, the new and powerful influence of neuroscience—both as neural modelling and as scanning technologies—reshaped the field by adding vast new territories to it. The brain was no longer a black box and decision-making could now become the object of more or less direct observation.

However, seeing something and understanding it may be two quite different things. When we marvel at real-time fMRI and other imaging, and establish statistical links among brain area activations, we only begin to realize how modest our intellectual grasp over human cognition is, and how much more theoretical our approach must be. That is why some of the contributions of mathematical and computational neuroscience that deal with human decision-making take centre stage in this book. They occupy its entire Part III, but the ground for them is prepared by the preceding two parts.

Part I carefully examines the merits and deficiencies of the classical utility paradigm, outlined in Chaps. 1 and 2. It begins with Daniel Bernoulli's seminal idea about subjective utility as a factor, influencing human motivation and choices. Further, Chap. 1 provides a concise discussion on the methodology of science in general, and explains the nature of scientific concepts, measurements, axioms, models and theories, all of them outlined in an accessible manner. The aim here is to facilitate the reader's orientation in the interdisciplinary territory of decision science.

Chapter 2 presents in sufficient detail the accomplishments as well as the weaknesses of the von Neumann–Morgenstern axiomatic system for economic rationality. Another highlight in the exposition is the Arrow–Pratt formula for individual risk-taking attitudes. The chapter closes with a twenty-first century review of the most important rationality principles, as they were ably examined by Busemeyer, Rieskamp and Mellers. By the end of Part I it becomes clear that a chain of events where, time and again, a new paradox of "irrational" behaviour is resolved by a new theory, which is soon challenged by yet another paradoxical finding, is not much of a success story. It must be recognized that deeper methodological obstacles exist that prevent universal solutions.

Part II begins with new hopes, raised by the foray of psychology and cognitive science into some of the outstanding problems in decision-making. The entire Chap. 3 is dedicated to the milestone contributions to the field by Tversky and Kahneman as seen from the standpoint of a scientist with a twenty-first century perspective. Chapter 4 outlines the Griffiths–Tenenbaum experiments that established exactly how people are optimal Bayesian statisticians in everyday judgements. The mastery of that study outlines some of the limits of the traditional—non-neural—mathematical modelling in cognitive science.

Finally, the entire Part III is dedicated to mathematical and computational neuroscience. A particular emphasis is put on the theories of the field's most prominent leader, Stephen Grossberg and his school of thought, somewhat at the expense of others. The approach is explored both theoretically and with experimental work. It begins with Chap. 5, which outlines the foundations of neuromodelling and clarifies its importance for decision-making. There, the Grossberg–Schmajuk theory of classical and operant conditioning and the Grossberg–Gutowski theory of cognitive-emotional interactions, together with their workhorse—the recurrent gated dipole—are summoned to explain how humans act when facing economic choices. In addition, some of the established concepts of the more traditional decision analysis are examined in the light of computational neuroscience. The outcome of this cross-paradigm exercise is a number of conclusions about the relevance of old concepts for new scientific approaches.

Chapter 6 presents an experiment on human intuition when doing economic choices. That study is guided by the theories outlined in the preceding chapter and shows how they can be harnessed in a practical application. Here, scientific details about the psychological, economic and computational issues abound in the exposition for two reasons. First, mathematical neuroscience is a young discipline, which is mainly occupied with discovering new knowledge, while using existing knowledge to guide new experimental studies has been rare. Thus, the limits within which neural models can explain empirical data from people's actions in different contexts are mostly untested. Such applications comprise a specific form of feedback and face challenges of their own. Chapter 6 seeks to contribute to their better understanding and to suggest ways to overcome some of them. The second reason for the highly technical discussion in this part of the book involves the very interesting effects related to human intuition that were discovered. However, they can be fully appreciated only in connection with the research methods that produced them.

The book's final Chap. 7 outlines a detailed analogy between the operations in an ART neural network taking place on the millisecond-to-second timescale, and some events and processes in hierarchical social organizations developing over months and years. This fractal-type analogy, alongside some others, forms the basis of a vision about a new kind of social science that could emerge by extending neuroscience modelling to the socioeconomic domain.

As the book is intended for advanced undergraduates and graduates, it had to convey major ideas in a way that would be as approachable as possible. To this end, formulae are kept simple. This meant deriving them in full when it was considered worthwhile, or skipping them entirely when it was justified. The choice in any particular case was guided mostly by a thought about the reader's convenience, but inevitably might have included the author's personal bias.

A bias of much greater proportions must be admitted here. Any writer would have struggled with the task of putting in such a book all the important achievements of decision science. This is simply not possible. Still, I am a little embarrassed that a number of distinguished scholars are just barely mentioned, and the names of many more had to be left out altogether. It is a consolation that a lot of other volumes have extensively presented and discussed what could not be accommodated here. By the same token, here I keep quiet about group decisions, game theory, operations research and other related fields. My topic is the scientific analysis of decision-making as practiced by the individual mind.

If lucky, decision science books sometimes attract the attention of readers from a wider audience besides the researchers in the narrow field. This makes sense because however technical one's profession or job might be, one must interact with other human beings whose choices inevitably affect him/her. Therefore, some knowledge about how people around us take their decisions is always helpful.

The book is intended to be read not only by decision scientists, but also by engineers, computer scientists, software developers, and the likes. These professions are behind some of the most spectacular intellectual triumphs of our time, such as building machines that autonomously explored other planets, beat the world champion in chess, and attained human-level mastery of the Atari computer games.

This successful trend is likely to continue with the advancement of new paradigms such as cognitive computing and of new temptations going along with

Acknowledgments

An author of a book is influenced by many people. Acknowledging them all by name is impossible, which makes writing this section always awkward. Yet I must express my gratitude to at least some of them.

Professor Dr. George Sotirov, my doctoral thesis adviser at TU–Sofia, and Prof. Dr. Arthur Markman from The University of Texas at Austin attracted my attention to the field of decision science.

Professor D.Sc. Irina Zinovieva from Sofia University St. Kliment Ohridski, Academician Vasil Sgurev from the Bulgarian Academy of Sciences, Academician Mincho Hadjiiski from the Bulgarian Academy of Sciences, Prof. Dr. Werner Güth from the Max Plank Institute of Economics in Jena, Prof. Dr. Henrik Egbert, and Prof. Dr. Nadeem Naqvi were my most helpful critics.

I benefited greatly from the discussions with Prof. Dr. Stephen Grossberg over the years.

Academician Janusz Kacprzyk from the Polish Academy of Sciences and Prof. D.Sc. Krassimir Atanassov–Corresponding Member of the Bulgarian Academy of Sciences, encouraged and supported the creation of this book.

A number of young colleagues at Sofia University St. Kliment Ohridski helped to design and conduct various psychological, economic and computational experiments. They were Stefan Pulov, Prof. Dr. Kalin Georgiev, Prof. Dr. Trifon Trifonov, Svetla Nikolova, Nikolay Georgiev, Dr. Anton Gerunov and Elitsa Hristova.

Professor D.Sc. Mirena Slavova from Sofia University St. Kliment Ohridski suggested the Latin terms Homo Aequanimus and omnium bonum.

Contents

Nomenclature

$A, B, C, \ldots$	Alternatives (prospects) containing risk
$A1, A2, B1, B2, \ldots$	
$A_I, A_{II}, B_I, B_{II}, \ldots$	
a_1, a_2, a_3	Constants in neural differential equations
b, b_1, b_2	
c, c_1, c_2	
$\mathbf{D}$	Weighted sum of decision factors
d_1, d_2, d_3	Weights in a decision rule
K	Winning supplier in the omnium bonum experiment
m, N, n	Number
p	Probability
P_a	Asking price
t	Time, or interim point in the accumulation of some stuff
T	Duration of a process or sum of a quantity
T^*	Median of T
u	Utility of a single outcome. In addition, u is used as equivalent to U in the integrands in Eqs. (2.6), (2.7), and (2.10)
U	Utility of an entire risk-containing alternative
U_t	Utility at time t
V	Utility function derived from a neurobiological model
w, w^+, w^-	Decision weighting functions
W, W_0, W_f	Total wealth, initial wealth, final wealth
x, x_i	Outcome in a prospect; any variable
y_i	Neuron activity
z_i	Neurotransmitter medium-term memory

z_{ij}	Long-term memory
ΔP	Difference between asking price and final price
$\pi(p)$	Subjective probability; probability weight in prospect theory
π_i, π_i^+, π_i^-	Decision weights in cumulative prospect theory

Part I
Subjective Utility

Chapter 1
Levels of Decision Analysis

1.1 The Idea of Subjective Utility

In the year 1731, Daniel Bernoulli, a Dutchman educated at the University of Basel, delivered a lecture at a meeting of the St Petersburg Academy of Sciences in Russia, later published as a *memoir* in the Academy Proceedings (Bernoulli 1738, 1954). Dealing with a seemingly minor and specific problem, in that work he did no less than lay the foundations of what would become *analysis of people's decisions*, an undertaking to grow in influence during the next three centuries. Bernoulli's main idea was simple yet intricate: for an individual, the utility of money and other material resources does not grow in proportion to the amount received, because, as personal wealth increases, one generally becomes less sensitive to any new acquisition. Thus, a gain is subjective and brings benefit depending much on the wealth one had beforehand. As Bernoulli put it,

> The price of the item... is equal for everyone; the utility, however, is dependent on the particular circumstances of the person making the estimate... A gain of one thousand ducats is more significant to a pauper than to a rich man though both gain the same amount.

In fact, Bernoulli invented the first mathematical model relating money to the way it is perceived. In his view, the *utility* of a small financial gain dW is proportional to the gain itself, and inversely proportional to the entire wealth W of a person. Therefore, the small gain's utility is $a.dW/W$, where a is a proportionality constant, $a > 0$. If someone's wealth was W_0 before receiving a sum, small or large, and rose to W_f afterwards ($0 < W_0 < W_f$), then the specific gain $W_f - W_0$ brought them utility u of this type:

$$u(W_0, W_f) = \int_{W_0}^{W_f} \frac{a.dW}{W} = a \ln \frac{W_f}{W_0}. \tag{1.1}$$

© Springer-Verlag Berlin Heidelberg 2015
G. Mengov, *Decision Science: A Human-Oriented Perspective*,
Intelligent Systems Reference Library 89, DOI 10.1007/978-3-662-47122-7_1

Fig. 1.1 Daniel Bernoulli's
utility curve. His analysis
accounted for only the
positive utility

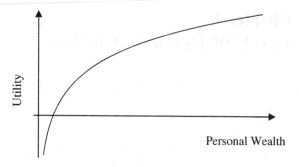

The greater the personal wealth then, the smaller the pleasure one receives from a particular gain, according to a logarithmic law (Fig. 1.1).

Bernoulli reflected on some pragmatic aspects of his postulate and gave an example with a hypothetical merchant hesitating to insure his ships in the North Sea. On one hand, seafaring trade is dangerous and about five out of a hundred vessels disappear, but on the other, all insurance companies demand too much for their services. A way of deciding would be to take into account the merchant's personal wealth. Were he rich enough, he could afford not to insure the goods and ships. Less well to do, he would be wise to protect against a potential loss. This analysis did not miss the insurers' point of view and business interest. In the memoir, a numerical example suggested how wealthy they should be, to afford certain financial deals.

Bernoulli's work seems to be historically the first to quantify an entity of subjective nature. Before him, mathematicians like Cardano, Pascal, and Fermat had already discovered ways to calculate risks in gambling using probability theory. Unlike all of them, however, Bernoulli proposed a way of accounting analytically for a solely human thing—the subjective utility, as the person taking a decision would perceive it (Bernstein 1996, 1998).

This achievement initiated research activities, which today belong mostly to two scientific fields. The first one is interdisciplinary and is called *decision analysis*. It deals with the methodology and mathematical methods involved in forming expert opinions and taking principled, that is, uncontroversial, decisions primarily for large-scale technical, ecological, financial, and other issues viewed as important by society. The second area is *microeconomics*—the study of the firm and its principle agent Homo Economicus, the decision maker maximizing utility under specific constraints. Both fields share almost identical conceptual framework where the utility function, its maximizing, risk avoidance etc. play the most important roles.

This book pays due respect to both fields, but its spirit belongs to neither. Alongside established results, rooted mostly in the classical utility theory, here I outline two other approaches to understanding how people make choices. The first comprises some of the most important ideas from the psychological research on decision-making over the last couple of decades. This part has a lot to do with economic psychology and behavioural economics. The second approach is less popular but, in my view, more promising. This is the approach of mathematical and computational neuroscience, and it deals with models that encompass the neurobiological facts about human thought and emotions relevant in making choices. My main objective has been to develop an integrated view of decision-making, and thus put in a new perspective people's actions in general, and their expert, technical, economic, and other choices in particular.

A useful starting point for this undertaking is *expected utility theory* and the contributions of its founders and later prominent figures John von Neumann, Oscar Morgenstern, Leonard Savage, John Pratt and others. There is a distinction between that theory and the *theory of marginal utility*, developed by the 19th century economists William Stanley Jevons, Carl Menger, and John Bates Clark (Galbraith 1987). Both theories share the seminal idea of diminishing marginal utility of a good, but make somewhat different use of it.

Box 1.1. The St. Petersburg Paradox

Daniel Bernoulli was inspired for his memoir by a scientific problem, formulated by his cousin Nicolaus Bernoulli, a mathematician and law professor. Nicolaus considered a game in which a coin is tossed until heads comes up. If that happens in the first toss, the player receives one ducat and the game ends. If heads comes up in the second round, the prize will be two ducats and the game will end at that point. Three rounds would mean four ducats, and four rounds—eight ducats, etc., in a geometric sequence. The player must name a price for which he would be willing to sell his right to play.

Bearing in mind the probabilities for heads in exactly the first, second, third round etc., which are 1/2, 1/4, 1/8, ... respectively, the gain g can be computed by the mathematical expectation:

$$E(g) = \sum_{i=1}^{\infty} p_i x_i = \frac{1}{2}1 + \frac{1}{4}2 + \frac{1}{8}4 + \frac{1}{16}8 + \ldots = \infty.$$

In the above equation x_i is the gain in ducats and p_i is the probability for heads in the i-th round after a sequence of $i - 1$ tales.

As Nicolaus Bernoulli remarked in a letter to his cousin, "Although the standard calculation shows that the value ... is infinitely great, it has to be admitted that any fairly reasonable man would sell his chance, with great pleasure, for twenty ducats" (Bernoulli 1738, 1954).

It has been estimated that in the beginning of 18th century a ducat could buy goods for approximately as much as US $40 in the last decade of 20th century (Bernstein 1996, 1998). The problem remained in the history of science under the name "St. Petersburg Paradox".

As it often happens, a profound new idea may be born at about the same time by more than one mind. Daniel Bernoulli admitted in his memoir that his cousin Nicolaus had written to him in 1728 that the mathematician Gabriel

Cramer had achieved identical results to Bernoulli's. In the latter's own words, "[...] Cramer, had developed a theory on the same subject several years before I produced my paper. Indeed I have found his theory so similar to mine that it seems miraculous that we independently reached such close agreement on this sort of subject." (Bernoulli 1738, 1954).

Decision-making under risk may be defined as a choice among *alternatives*, or random events to happen in the future, with explicitly stated outcomes and *known* probabilities. When vagueness is even greater and the probabilities are unknown, this situation is scientifically termed *uncertainty*. Often alternatives can be represented by random variables with discrete distribution laws. For example, a finalist in a tennis tournament would receive an amount of money x_1 should he win; should he loose, he would get only x_2, where $x_1 > x_2$. But what are the odds for any of these to happen? Imagine that he has already played with his rival n matches and has won n_1 of them $(n_1 \leq n)$. Therefore, one way to assess his probability of winning this time is $p = n_1/n$. Formally, his position can be described as $(x_1, p; x_2, 1 - p)$. We can now introduce the following general.

Definition An *alternative* $A \equiv (x_1, p_1; \ldots; x_n, p_n)$ is a set of outcomes $x_1, \ldots, x_n$ and their probabilities of occurrence $p_1, \ldots, p_n$, whereby $p_1 + p_2 + \ldots + p_n = 1$.

The outcomes x_i where $i = 1, \ldots, n$ are money or another kind of resources. Note that the alternative is the entire set of outcomes and probabilities, and not each of them. Throughout this book, as a synonym to alternative will be used also the term *prospect*.

When an individual chose an alternative, they are said to have made a contract, which could be formal (in the legal sense), or just psychological—in the sense that they accepted and appreciated the potential consequences. The above definition is fairly general and can adequately describe alternatives in insurance, banking, stock trading, and many other businesses. One way of stating the outcome of a choice would be, as Nicolaus Bernoulli did (see Box 1.1), by using the mathematical expectation:

$$E(x_1, p_1; \ldots; x_n, p_n) = p_1 x_1 + \ldots + p_n x_n. \tag{1.2}$$

When Daniel Bernoulli invented the *expected utility u,* he in effect introduced in the scientific analysis the standpoint of the individual. Thus, for the tennis player, as well as in many other cases it became possible to compute the subjective utility U of the entire future situation, i.e., the alternative $(x_1, p_1; \ldots; x_n, p_n)$, by using the utility functions $u(x_i)$ of the single outcomes:

$$U(x_1, p_1; \ldots; x_n, p_n) = p_1 u(x_1) + \ldots + p_n u(x_n). \tag{1.3}$$

In the famous memoir, Eq. (1.3) defined "moral" (i.e. psychological) expectation of an alternative, which is different from its mathematical expectation.

Introducing the individual perspective as a subject of scientific research was such a giant step that it remained unutilized for many decades. As late, as 1854, the German economist Hermann Heinrich Gossen rediscovered roughly the same idea by stating that when a good is consumed to satisfy a need, moment by moment the need would diminish, eventually coming to nil. A thirsty man drinks a cup of water with great pleasure, the second cup less so, and the fifth cup becomes useless. Thus, Gossen's satisfied need was in fact Bernoulli's subjective utility, perhaps with a bit of emphasis on the act of consumption. Contemporary economics recognizes this as the First Law of Gossen.

Yet, more than a century after its original inception, circumstances had still not been ripe for that discovery; Gossen's Law remained disregarded by his contemporaries just as Bernoulli's ideas. Advancement in other fields was necessary, before the insight of both men could be usefully integrated in science. In 1834, the German physiologist Ernst Weber studied human physiological and psychological reactions to stimuli of varying physical intensity. He noticed that there existed a minimum magnitude of the stimulus, a threshold, below which it got impossible to perceive, and gave it the name Just Noticeable Difference, JND. As the stimulus intensified, the threshold also grew, with the ratio between them staying roughly constant. Interestingly, the different senses like sight, hearing, etc. have different JNDs. For example, a change in light brightness must be at least 1.7 % to be noticed by the human eye (Schiffman 1976). Around 1860 another German physiologist, Gustav Fechner, continued Weber's work and conducted similar investigations. He concluded that the intensity of the sensory reaction is proportional to the logarithm of the stimulus intensity. Today the effect is known as the Weber-Fechner Law and graphically looks exactly like the curve in Fig. 1.1 above the abscissa.

There is a deep relation between how people react to physical stimuli and to influences of more abstract nature. For example, the effects of financial changes in one's life are felt in a similar way to physical impacts, as was established in 20th century by the economist Allais (1953), and the psychologists Kahneman and Tversky (1979, 1984, 2000; Tversky and Kahneman (1981, 1986). The latter two have concluded that the brain processes in identical way changes in the physical environment and changes related to wealth, income, and prestige. This idea has had some unexpected consequences, one of them being the question how financial income and psychological wellbeing are related.

In an international survey, social scientists Inglehart and Rabier (1985) studied general life satisfaction in 19 developed countries in relation with the average annual income per capita. Their data are presented in Fig. 1.2 and show a very small increase of the satisfaction in connection with the rise of wealth. Indeed, the two poorest countries in the sample, Greece and Spain, had an average income of less than US $2900, and showed happiness around six points in the zero-to-ten scale. At the same time, the peoples of the richest countries in the sample, Sweden and USA, received on average more than US $10,000, and were only a tiny bit happier; in fact, the Spaniards were only a point behind the Americans. That single point came along with approximately four times larger income. Often such graphs are plotted

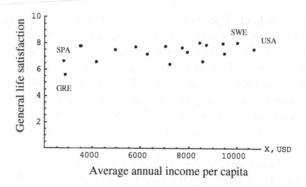

Fig. 1.2 Plot of general life satisfaction versus income in USA, Japan, and 17 West European countries (Inglehart and Rabier 1985). *Notation SPA* Spain, *GRE* Greece, *SWE* Sweden

with the income-per-capita variable in logarithmic scale, just to highlight how little happiness and income correlate.

These issues have been first investigated by Richard Easterlin and his colleagues (1974, 2010) with data from developed, developing, and East European countries. They found that in the long run, happiness does not increase as a country's national income per person rises. This is in contrast with data showing that within a given country people with higher income are more likely to be happier—a phenomenon known as the Easterlin Paradox. Today the standard explanation is that happiness rises together with income until the basic human needs are satisfied, but no further. Beyond that point, one begins to compare one's material condition with the well-being of others, and that increasingly forms one's level of reference.

In the 21st century, this field developed further and defined a "happiness function" as the ability to transform objective circumstances into subjective well-being. In particular, the distinction was made between "life satisfaction", which has a predominantly cognitive content, and "happiness", which is a mix of cognitive and emotional elements (Helliwell et al. 2013), although both are reported to be highly correlated (Clark and Senik 2012).

Comparisons among countries have produced intriguing findings (Inglehart et al. 2008; Sacks et al. 2010). For example, a detailed econometric study by Senik (2014) showed that the relative unhappiness of the French people in comparison to other nations with similar level of development and income such as Belgium, Great Britain, Canada, Austria, and The Netherlands is due in part to cultural factors. That conclusion was based on statistically significant coefficients in a regression analysis and unfortunately could not say more about the exact nature of those factors.

For another example, the Sacks et al. (2010) study showed Bulgarians to be a little wealthier than Brazilians, but twice as gloomier, a fact that provoked some witticisms by *The Economist* (2010). Part of the explanation for this particular national unhappiness might be that Bulgarians tend to compare themselves with West Europeans whose living standards have always been higher. A closer look at the issue is possible due to a representative survey on the quality of working life

Fig. 1.3 Link between satisfaction from income, and general job satisfaction in Bulgaria in 1994

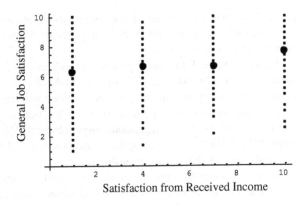

conducted among 1104 employed respondents in Bulgaria in 1994–1995 (Roe et al. 2000; Zinovieva 1998). Figure 1.3 plots their general job satisfaction, a subset of life satisfaction, against their contentment with the received income. The small dots designate answers of at least one respondent, and the large dots show averages at each level of income satisfaction.

Apparently, there was a large diversity of attitudes, as extremely satisfied and extremely unhappy people existed regardless of their ability to make ends meet. Down to the right in the plot, individuals disliked their job and kept it only for the money, while people in the upper left corner thoroughly enjoyed what they did, notwithstanding their miserable income. Overall, the big dots suggest a slight tendency of increasing job satisfaction alongside increasing income contentment. The large dispersion of responses indicates that the entire picture is far from simple as many other factors may be influencing job (and life) satisfaction.

1.2 Descriptive and Normative Theories

Throughout the book, we will come back to the question what motivates people to make certain decisions. Naturally, the next thing to ask would be, was a choice optimal. Thus, decision analysis has two aspects. Usually people react intuitively and their actions are frequently less than perfect. There is the need to know then, what has influenced them and what mental mechanisms have been at work. On the other hand, to know what could have been the best solution would mean to be armed with formal methods, procedures, and algorithms to compute it.

Therefore, it is natural that two different types of theories would seek to answer these two questions. Psychological theories, sometimes very mathematical, try to explain the observed behaviour. They are termed *descriptive* as they characterize, or describe what people do, which does not exclude sometimes attempts at prediction. In contrast, there exist other theories, often even more mathematical, which are called *normative* because their methods aim at finding the best solution in a formal sense. This division is obvious today, but it has not always been so. Overall, science went through a long evolution involving certain controversies, until it got to recognize the two approaches.

Thinkers have formulated postulates and axioms, trying to summarize the ingredients of a "good decision", only to find out how shaky their theorizing had been. Almost every new theory held the ambition to overcome the deficiencies of its predecessors and to unite the normative and descriptive aspects in one final perfect edifice. Soon, however, ingenuous psychological observations or experiments produced puzzling results, comprising paradoxes that defied the wisdom and even the credibility of the dominant postulates.

To pay respect to some theoreticians' efforts—consisting sometimes of little more than certain beauty of mathematical ideas—the theory in question would be rebranded "normative", a polite way to say that it is no longer considered all-encompassing. Simply put, it would be officially recognized as offering a less than accurate account of the observed behaviour, but—should people choose to follow it anyway, and abide by certain formal rules—that theory will guarantee an optimal result in some analytical sense. Then, the door would be left open for the next candidate for a general theory of decision-making.

To understand why such developments are possible and happen in practice, one must know something about the nature of scientific enterprise and its machinery, which contains concepts like postulates, axioms, models and other related notions. In short, one needs some basic understanding of the philosophy of science, which is the subject of the next section.

1.3 Constructs, Definitions, and Scientific Measurement

Although each scientific discipline has its own tools and methods, these are all based on principles, common to all sciences. Figure 1.4 shows a diagram of the different elements that build up a scientific theory in any field of knowledge (Torgerson 1958). The shaded area to the right represents the observable world we explore. Our goal is to understand and explain its phenomena and, if possible, to be able to predict them.

The circles to the left represent the theoretical constructs in each discipline. For example, in various branches of physics that can be length, force, electric current etc. In every science, it is necessary that at least some of the theoretical constructs be defined so as to have a clear connection with phenomena and objects from the real world. That is how a science can be linked to empirical observations and can become useful in practical terms. In Fig. 1.4 these are the constructs C_1', C_2', and C_3'. Their definitions, shown with double lines, contain a prescription of actions that always yield the same result when conducted by an expert. The sequence of these operations and procedures makes up the operational definition, also called rule of correspondence, and shows how exactly a scientific theory connects with the real world.

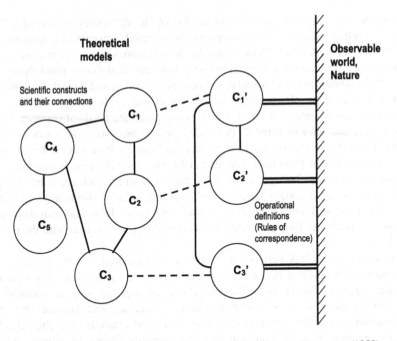

Fig. 1.4 A diagram illustrating the structure of a scientific field. After Torgerson (1958)

Many constructs are defined only theoretically and in Fig. 1.4 these are $C_1, \ldots C_5$. Together with the connections between them, they form theoretical models. Purely theoretical models exist perhaps only in mathematics and in philosophy. Without operational definitions, they can be analyzed only in terms of their logical consistency. Their lack of connection with natural phenomena has led some mathematicians to state that in mathematics no discoveries are made, but only statements (propositions) are proven. Others, however, have considered the word "discovery" to be too nice and elegant, and have insisted on its broader understanding, encompassing also the intellectual achievements of mathematicians.

In addition, even that most abstract science can sometimes use rules of correspondence. For example, there is both a theoretical definition of the concept *probability*, due to the Kolmogorov axioms, and an operational one—the frequency of occurrence of an event in a long sequence of trials. We used the latter to define the chances of the tennis player in his final match of the tournament. This is the *statistical probability*, also called *frequentist probability*. Decision analysis makes extensive use of it.

When a branch of science contains theoretical models of connected constructs, and at the same time for some of them exist rules of correspondence, then there is a theory that can be empirically verified. Eventually, such theories are refined and become very useful in explaining and predicting phenomena from the surrounding world. Consequently, it becomes possible to invent and then manage artificial systems. For example, people's knowledge of electricity and magnetism led to the creation of electric power stations and the use of electric energy in everyday life. Similarly, semiconductor physics is among the essential elements of contemporary computers and hence, of internet and the information society. In that sense, a scientist once observed that there is nothing more practical than a good theory.

With the advancement of science, its constructs develop and their definitions get refined. Medieval people measured length in feet and yards (a sequence of three feet), apparently slightly different for each person, until the needs of commerce and governance, and later of science, demanded greater precision and a stricter definition. Towards the end of the 18th century, in France a base unit of length, one *metre*, was established and then redefined eight times over the next two centuries to improve its accuracy. Today it is related to the length of the path travelled by light in vacuum during a certain time interval. Another example from physics is the definition of *power* and its measurement unit *horsepower*. Its operational definition has also evolved over time and today has no connection whatsoever with the power of a horse.

There is a direct link between the stability of definitions and the precision of measurement. Both lie at the heart of people's ability to use theoretical models for practical purposes. In our time, the natural and engineering sciences seem to be ahead of the social sciences in both measurement accuracy and construct stability. The economics variable *inflation* can serve as a good example. It is theoretically defined by Irvin Fisher's equation from the quantity theory of money, which operates with variables like price level, total stock of money, velocity of money circulation etc. On the other hand, an operational definition of inflation, i.e. a method for its measurement, is the consumer price index (CPI) that rests on a basket of consumer goods—a set of commonly bought household goods and essential services. Apparently, the latter definition is destined to change over time, unlike its theoretical counterpart. If nothing else, the basket will evolve due to social progress and technological innovation bringing out new products in the market. One can view the circles C_1' and C_1 in Fig. 1.4 as an illustration of the two definitions, with the dotted line indicating the somewhat unsteady relationship between them.

The stronger the unity between theoretical and operational definitions in a scientific field, the greater is the researchers' capability to grasp theoretically and then predict important phenomena. In that sense, it has been noted that each science is identical with the instruments and methods for measurement that it commands (Kline 1985). Decision science, the subject of this book, is not a unitary science but an interdisciplinary one, and offers numerous examples of how true that statement is.

1.4 Postulates and Axioms

A main objective of decision analysis is to formulate a set of principles, or postulates, that describe adequately human choice under the widest possible variety of situations. It is desirable that even some of the essential elements of people's motivation be encompassed by those postulates. A prominent example is the principle of utility maximizing that is shared both by expected utility theory and by the theory of marginal utility. It views the individual as Homo Economicus— someone who is well aware of their own needs, and the opportunities to satisfy them as made available by the surrounding physical and social environment. She or he selects the best for themselves within the limits of disposable income, health considerations, and ethical and aesthetic values. That postulate—insightful at the time of its inception and elaboration in the writings of John Stuart Mill, Adam Smith, David Ricardo, Wilfredo Pareto and other prominent figures—has been challenged a lot over time.

Some postulates may be formulated mathematically and take the shape of a system of axioms. Any such system must meet three requirements: consistency, completeness, and independence of the axioms in it. It may happen though, as it certainly has with the von Neumann & Morgenstern (1944, 10.2) expected utility theory, that completeness is not satisfied. This simply means that outside of mathematics, the other sciences may adopt a less strict view. As we will mention axioms throughout the book, it is worthwhile to say something about what they really are.

The ancient Greeks believed that an axiom is a self-evident premise that should simply be accepted as true. Modern philosophy of science has moved further from that position. Our contemporary understanding is due to the French mathematician from the late 19th and early 20th century Henri Poincaré, who stated that axioms could be neither "a prori judgements", nor "experimental facts". In his view, axioms,

> ...are [...] conventions; our choice among all possible conventions is guided by experimental facts; but it remains free and is limited only by the necessity of avoiding all contradiction. Thus, it is that the postulates can remain rigorously true even though the experimental laws which have determined their adoption are only approximative. In other words, the axioms [...] are merely disguised definitions. (Poincaré 1902).

Further, Poincaré believed that no geometry could be more correct, or true, than any other, but can only be more *convenient*. Apparently, this statement can carry over to many other branches of science. Throughout the present book, we adopt exactly that view—axioms and postulates are conventions, concisely summarizing human intuition and experience. The situation with all axiomatic systems in decision analysis—a discipline using mathematics but itself not part of mathematics—is exactly as described. In addition, John von Neumann and Oscar Morgenstern claimed that an axiom must have "intuitive meaning" and is not "purely objective". Moreover, they offered, in their own words, "a set of axioms which seems to be essentially *satisfactory*" (von Neumann and Morgenstern 1944, 3.5.2). Modern economics assumes that the axioms do no more and no less than summarize the empirical content of a theory (Gul and Pesendorfer 2008).

The theories in economics and decision analysis are mathematical structures employing idealizations, governed in large measure by postulates, axioms, and definitions. Often, the simplifications introduced in this way may not cause a significant departure from reality. However, each attempt at practical application raises the question whether the object of study and its idealization are close enough (Kline 1985). The answer may be specific for each case, yet two points seem to be general.

First, there is the factor of the scientists' ingenuity—their ability to create useful new definitions, theoretical and operational, and even "disguised", that is, axioms. The latter should be "convenient" and "satisfactory" enough, to be able to further significantly our knowledge. Secondly, the technological and philosophical achievements of every age determine what instruments for observation and measurement are available, and inevitably restrain human imagination in its quest for new and more productive operational definitions in particular and scientific constructs in general. But then come new generations who take on the torch, face new obstacles, come up with new solutions, and ultimately open up new horizons.

1.5 Three Levels of Decision Analysis

The classical method for advancement in many research fields—by observing and experimenting, then formulating postulates or first principles, and then developing models and theories based on them—proved productive in the natural and engineering sciences, but faced substantial difficulties with regard to human decision-making. One reason seems to be the enormous complexity of the brain and the way it operates—in fact, understanding *that* stands as perhaps the greatest challenge before the researchers of our time. The century-long play of normative and descriptive theories has produced results deserving respect for both the sophisticated mathematical constructions, occasionally very useful in practice, and the psychological insights that have augmented our knowledge of ourselves as human beings.

However, in our time the state of the field is essentially disparate and no unifying paradigm is in sight. There exist plenty of axiomatic systems seeking to grasp human choice and achieving only partial success. Yet, researchers seem not to be interested in identifying a common set of postulates, around which a consensus could be built, and only the von Neumann–Morgenstern theory stands as a shared starting point. An optimist's way of looking at the situation would suggest that decision analysis is vigorously developing and that this is probably the best to be desired. A different perspective may point to the natural sciences, where a common basis of fundamental principles has guided research towards many successful results, to the envy of many social scientists.

There is of course, truth in both positions and they may even be not too hard to reconcile. Part of the complication stems from the elusive nature of human choice as a subject for scientific analysis. It is indeed difficult to define what exactly constitutes a decision: Is it a purely philosophical concept? An act important only through its consequences? Or is it a continuous-time psychological process with firm biological roots? If we accept that it is all of these things at once, then we could have our hands free to tackle each facet separately, adopting the most suitable approach for every particular case. Then we may try to arrange the many research

---

1. Decision Analysis and Economics

Expected utility theory; Decision theory and risk analysis;

Rationality principles and paradoxes of irrationality

---

2. Psychology

Prospect theory and framing effects; Heuristics, biases, and intuition;

Fuzzy trace theory; Dual process theories;

Subjective statistics and Bayesian cognition

---

3. Neurobiology

Mathematical and computational neuroscience;

Brain imaging and neuroeconomics

Fig. 1.5 Levels of decision analysis

venues by what would emerge as their shared characteristics, and arrive at a picture of relative simplicity.

One such organizing effort is presented in Fig. 1.5 where I summarize the scientific venues of decision science as they have developed over decades and centuries. Historically the first was the line of research initiated by Daniel Bernoulli, later continued by a number of economists, and reaching maturity in the opening chapter of the famous von Neumann–Morgenstern book Game Theory and Economic Behaviour. That volume was planned as a foundation for a new universal economic theory, and indeed, it became very important for academic economists as well as for engineers, managers, financiers, and administrators, seeking scientific guidance to find the best solutions for vital societal problems.

I put all this diversity and long tradition in one "pot", or level of analysis, because of the similar goals they have pursued, as well as the research methods they have employed. Some of this level's key concepts were introduced in the previous sections, and the entire Chap. 2 is dedicated to it. However, we do not deal with game theory at all in this text, simply because its perspective is human-oriented, and we are interested in the actual behaviour of the agents, rather than in prescriptions how it could be optimized in the course of economic activity, which is what game theory is about. For very similar reasons operations research is not discuss either.

As decision analysis developed, it became increasingly obvious that its limitations could be overcome only by more knowledge about the decision maker.

The second half of the 20th century shaped the researchers' understanding that psychology is indispensable in that process. However, even before, progress was made most often when some insight about human nature was incorporated into the axiomatic treatments of the subject. A little-known fact is that Daniel Bernoulli himself graduated in both mathematics and medicine. Over the 20th century, people with such interests usually took psychology as the second major. Indeed, the latter science became a necessity simply because its experiments produced the paradoxes that normative theories failed to explain. Moreover, it managed to find some very constructive solutions to the problems it created, of which most notable was *prospect theory* of Amos Tversky and Daniel Kahneman. That theory combined formalized analytical with more traditional psychological methods and was able to explain a large variety of seemingly irrational instances of economic behaviour. Published in 1979 in *Econometrica*, it influenced generations of economists and other social scientists, and alongside its improved variant from 1992, it earned Daniel Kahneman the Nobel Prize in 2002 (Tversky had died some years earlier). Prospect theory helped researchers recognize how complex and unpredictable are the deep mental mechanisms that guide human choices.

The decision maker's mind possesses a deeply rooted intuitive toolbox, vital for coping with challenges that arise moment by moment, and this is the set of mechanisms known as cognitive *heuristics*. Unconsciously and almost instantaneously, they produce approximate estimates for important parameters of the surrounding environment or the social situation that guide us successfully out of serious danger or unpleasant embarrassment, or help us avoid them altogether. Smart and refined by evolution as they may be, these devices are not infallible—on rare occasions, they misinterpret reality and mislead us into wrong judgements and decisions. Moreover, they do so systematically, which amounts to cognitive *biases*. In science, such instances are very much loved by researchers, as they give them opportunities to apply witty methods, eventually documenting nature's "imperfection", or at least its suboptimal performance under laboratory conditions. The field of heuristics has witnessed an interesting dispute between Kahneman and Tversky—two of its founders, and Gerd Gigerenzer, another prominent figure in cognitive psychology. The former two showed ingeniously how heuristics could be fooled, while the latter insisted how useful they were. Apparently, the debate was more about where to put the emphasis rather than about a deep division on theoretical grounds. Naturally, we place the field of heuristics and biases alongside the other psychological domains in the second level of decision analysis in Fig. 1.5. Kahneman's 2011 masterpiece Thinking, Fast and Slow is the reference point for this field.

A related line of psychological research focused on the human ability to resort to experience and use it as *intuitive statistics*—also called *subjective statistics*—when adapting and reacting to important environmental events. Because the ordinary situations in life often have similar features, evolution has armed all living creatures with subtle mental mechanisms to develop memory for the common and the repeated. The relevant mathematical tool to study such phenomena is the Bayesian approach, which computes the probability of occurrence of an event by using the accumulated statistics for the set of related cases from the past. A breakthrough in this field was achieved in 2005, when Thomas Griffiths and Joshua Tenenbaum discovered exactly how humans are intuitive Bayesian statisticians in everyday life and what is the importance of this result. These findings are discussed in Chap. 4 of the book.

Much like the way psychology was vital for understanding the decision maker, the scientific community recognized that knowledge about the biological substrate is necessary to get even deeper into the fabric of the cognitive process. But how can we look inside one's head to see how decisions are made? This question had disheartened many philosophers for millennia, until 20th century technology finally provided the opportunity to do—if not exactly that, then something a lot like it. Certain achievements of engineering physics met the needs of medicine, and led to the development of the functional magnetic resonance imaging (fMRI)—a method for indirect measurement of activity in different brain areas via monitoring the change in the blood's magnetic properties. Exactly that is detected by the fMRI scanner, which then produces a map of the neuronal activation over substantial parts of the brain, especially its cortex. Used alone or complemented with other measurement and diagnostic methods, this technology offered unprecedented opportunities to study the activity of the human brain in real time. Psychologists got interested in conducting experiments with it to test their theories, and towards the end of the 20th century, they were probably more actively using it than the physicians were. In early 21st century, some economists joined the enterprise, intrigued both by the new fashionable thing, and by the prospect of investigating what really happens in the brain of a person making economic decisions. That is how "neuroeconomics" was born. Unrecognized by the mainstream economists as a territory belonging to their realm, nonetheless it developed vigorously, and firmly established itself as an important field of interdisciplinary research. Still, the brain-scanning technology continued to provoke a lot of skepticism as it was able to identify neither the activity of the single neuron, nor the mechanisms behind the orchestrated functioning of the different brain areas. As Russell Poldrack put it (Smith 2012), fMRI measures stuff that neurons are doing, that is relevant to mental function. But what exactly—it remained not quite clear.

While improving the methods for brain imaging, we should not forget the importance of developing theoretical models to explain what we have recorded. Einstein once noted that having plenty of observations cannot replace the need for a theory, and cannot even bring us closer to an explanation. Decision analysis has suffered from the chronic problem of not being able to develop a set of stable fundamental principles that can guide its further evolution.

The postulates of economics and even psychology have proven to be conceptually weak when it came to describing the whole variety of choice behaviour and certainly, they could not be convincingly related to the findings of brain imaging studies. In this book, I suggest that an intermediary layer of knowledge may become instrumental in finding those links.

That is how we arrive at the topic of the last part of the book, which is the neurobiological modelling of the decision-making process. In it, the centre stage is given to the dynamics of the cognitive interactions that produce adaptive and intelligent behaviour in humans (and other higher species). In a way similar to the natural sciences, this approach begins by examining the variety of empirical find-ings related to the investigated cognitive process. The researchers then seek to formulate theoretical principles about it, which are implemented afterwards in mathematical and computational models. The origins of this particular scientific endeavor lay in the works of Warren McCulloch and Walter Pitts, who resorted to mathematical logic in an attempt to formulate a set of axioms about the functioning of the nervous system, as it was known in the 1940s. In their time, not enough knowledge was available to make that effort sustainable, and future discoveries proved their axiomatic system unrealistic.

However, the potential of the neurobiological approach has never been in doubt, and it was vindicated a couple of decades later. Stephen Grossberg, a psychologist with a doctorate in mathematics, gave it a substantial boost by choosing an entirely new road for theoretical advancement. Instead of axioms, he introduced to the analysis of cognitive processes a set of differential equations. His method has been to study the available empirical observations about a phenomenon, then formulate postulates, and then identify the simplest structure of connected neurons and syn-apses that could implement those postulates. At that point, appropriate differential equations are introduced to describe the operation of the identified neural circuit. Next, the new model's properties are studied analytically and then by computer simulations, to validate its correspondence with the empirical data.

Naturally, the first such models dealt with simple interactions like those of a neuron receiving signals from other neurons or sending signals to another set of neurons. Later, models that are more complicated emerged, dealing for example with the simplest neural configuration capable of producing a primitive emotion. It turned out that a circuit of six neurons is enough to account for the balance of opposing emotions such as fear-relief, hunger-satiation etc. Along that road, a family of mathematical models emerged and described more sophisticated neural networks corresponding to more complex psychological phenomena like short- and long-term memory, pattern recognition, and eventually, complex cognitive-emotional inter-actions forming the fabric of observable behaviour. Since the 1980s, the achieve-ments of the Grossberg School have sporadically been tried in models of economic and other kinds of decision-making. Important work in that direction was carried out in the first decade of this century.

The wider discipline treated in that part of the book is often called mathematical neuroscience, or computational neuroscience. A less popular name is connectionist cognitive science. Including it in this book may surprise some traditionally minded experts in decision studies. However, it has the potential to resolve many of the issues arising within the social sciences with their fragmented axiomatic base. In addition, it can complement the classical psychological investigations, and in many instances can provide theoretical guidance for brain imaging studies.

Of course, this point of view is debatable. Objections may arise, for instance, against the attempt to organize all types of human decision analyses in a simple classification as shown in Fig. 1.5. Further, some might say that decision science belongs to engineering and operations research as much as to the social sciences, and that position would not be without justification. May be more importantly, one could ask why various branches of applied mathematics contributing to decision analysis are not represented in it. A criticism towards the psychological level of investigation may suggest that the latter include, as a minimum, also group decision-making with its obvious relations to social psychology and other psychological subdisciplines. Finally, putting on the same research map utility theory and computational neuroscience may seem unwarranted because, still a relatively small number of decision studies have benefited from the analytical techniques of the latter.

The fact that all these, and potentially many more objections could be raised highlights, among other things, the point that contemporary analysis of decisions is a multifaceted field that has still not reached maturity. It is inhabited by a variety of approaches with no apparent relationships among them, and even their proponents are sometimes unaware of each other's mere existence. In that situation, I believe that proposing a simple intuitive classification such as the one in Fig. 1.5 could do more good than harm. One useful thing that it says is that decisions have been studied at levels, matching a natural hierarchy of entities whereby the one above is an extension and sometimes a generalization of the collective behaviour of those at the preceding rung of the ladder. For example, the biological level deals with ensembles of neurons and synapses, but their individual actions aggregate and disappear in the melting pot, which comprise the operations of heuristics, biases, intuition etc., all of them belonging to psychology. Similarly, the social sciences at the top of the hierarchy have as primary focus either the collective behaviour of people, or the choice of the single individual once it was made, declared, and registered, rather than the multitude of psychological traits characterizing each person or the motives leading to a particular choice.

In other words, the approaches to decision studies have been grouped in such a way as to mirror nature. However, the scientific tools applied at each level are specific and differ radically from what works at the other levels. This may displease some advocates of theoretical unification (mostly coming from the natural sciences) who would wish to see a single edifice—an all-encompassing construction, capable of explaining each human decision as a sheer special case. Here, the view is taken that such an ambition is probably futile. By now, it should be obvious that "decision" is a cultural concept denoting different things in at least three separate contexts, addressed by three different levels of scientific inquiry. This is the position adopted in the book as it approaches the existing situation at each layer of analysis in the following chapters.

References

Allais, M. (1953). Le comportement de l'homme rationnel devant le risque: Critique des postulats et axiomes de l'ecole americaine. *Econometrica, 21*, 503–546.

Bernoulli, D. (1738, 1954). Specimen Theoriae Novae de Mensura Sortis. *Commentarii Academiae Scientiarum Imperialis Petropolitanae*, Tomus V, 1738, 175–192 (Exposition of a New Theory on the Measurement of Risk, 1954. *Econometrica, 22*, 23–36).

Bernstein, P. L. (1996, 1998). *Against the Gods: The remarkable story of risk*. New York: Wiley.

Clark, A., & Senik, C. (2012). Will GDP growth increase happiness in developing countries? In *Volume of the 8th AFD-EUDN Conference: Measure for Measure. How well do we measure development?* (pp. 99–176). STIN, Paris.

Easterlin, Richard A. (1974). Does economic growth improve the human lot? In Paul A. David & Melvin W. Reder (Eds.), *Nations and households in economic growth: Essays in honor of Moses Abramovitz.* New York: Academic Press Inc.

Easterlin, Richard A., McVey, Angelescu, McVey, Laura Angelescu, Switek, Malgorzata, Sawangfa, Onnicha, Zweig, Smith, & Jacqueline, (2010). The happiness-income paradox revisited. *Proceedings of the National Academy of Sciences of the United States of America, 107*(52), 22463–22468.

Galbraith, J. K. (1987). *A history of economics: The past as the present.* UK: Hamilton.

Gul, F., & Pesendorfer, W. (2008). The case for mindless economics. In A. Caplin & A. Shotter (Eds.), *The foundations of positive and normative economics.* Oxford: Oxford University Press.

Helliwell, J., Layard, R., & Sachs, J. (2013). *World Happiness Report.* United Nations.

Inglehart, R., & Rabier, J. R. (1985). If you're unhappy, this must be Belgium: Well-being around the world. *Public Opinion, 8,* 10–15.

Inglehart, R., Foa, R., Peterson, C., & Welzel, C. (2008). Development, freedom, and rising happiness: A global perspective (1981–2007). *Perspectives on Psychological Science, 3,* 264–285.

Kahneman, D., & Tversky, A. (1979). Prospect theory: An analysis of decision under risk. *Econometrica, 47*(2), 263–291.

Kahneman, D., & Tversky, A. (1984). Choices, values and frames. *American Psychologist, 39,* 341–350.

Kahneman, D., & Tversky, A. (Eds.). (2000). *Choices, values and frames.* New York: Cambridge University Press and the Russell Sage Foundation.

Kline, M. (1985). *Mathematics and the search for knowledge.* New York: Oxford University Press.

Poincare, H. (1902). *The foundations of science: Science and hypothesis* (English Trans. 1913). The Science Press, New York and Garrison, NY.

Roe, R. A., Zinovieva, I. L., Dienes, E., & Ten Horn, L. (2000). A comparison of work motivation in Bulgaria, Hungary, and the Netherlands: Test of a model. *Applied Psychology: An International Review, 49*(4), 658–687.

Sacks, D. W., Stevenson, B., & Wolfers, J. (2010). Subjective well-being, income, economic development, and growth. *NBER Working Paper* 16441.

Schiffman, H. R. (1976). *Sensation and perception: An integrated approach.* New York: Wiley.

Senik, C. (2014). The French unhappiness puzzle: The cultural dimension of happiness. *Journal of Economic Behaviour and Organization, 106,* 379–401.

Smith, K. (2012). Brain imaging: fMRI 2.0. *Nature, 484,* 24–26.

The Economist. (2010). *The Rich, the Poor, and Bulgaria.* December 16th, 2010.

Tversky, A., & Kahneman, D. (1981). The framing of decisions and the psychology of choice. *Science, 211,* 453–458.

Tversky, A., & Kahneman, D. (1986). Rational choice and the framing of decisions. *Journal of Business, 59*(4), 251–278.

Torgerson, W. (1958). *Theory and methods of scaling.* New York: Wiley.

von Neumann, J., & Morgenstern, O. (1944, 1947, 1953). *Theory of games and economic behaviour.* Princeton: Princeton University Press.

Zinovieva, I. L. (1998). *Why do people work if they are not paid? An example from Eastern Europe.* William Davidson Institute Working Paper No 206, University of Michigan.

Chapter 2
Utility and Rationality

2.1 The von Neumann–Morgenstern View on Economic Motivation

As we saw in the previous chapter, the link between income and wellbeing is complicated because they correlate only until the basic human needs are satisfied. Modern economic psychology has discovered that beyond that level, money changes its psychological function and becomes means for participation in social relationships, as well as for managing, and perhaps reformulating, one's personal identity within those relationships (Ahuvia 2008).

For the economist seeking to understand what motivates people in their choices there is too much psychology in that. More appreciated would be a simpler model that accounts for some essential aspects of human motivation and ignores everything else. The methodological ways to do so are endless and naturally, some are more promising than others are. Economics and decision science have a consensus that a very important achievement in that respect is the von Neumann–Morgenstern theory of expected utility (von Neumann and Morgenstern 1944, 1953). Over the decades after its publication, it has been upgraded and refined by thousands of scholars and this activity continues even in the 21st century. At the same time, a growing number of instances emerged that clearly showed how the theory was unable to explain real economic behaviour. Today the limits of its validity are well established. Although it gradually gives way to new theories integrating relevant psychological knowledge, it remains a cornerstone in the development of science. Historic respect as well as necessity leads us to discuss it briefly here.

Characterizing the economic agents, John von Neumann and Oscar Morgenstern took a position that was typical for their time, and in some sense remains widely adopted even nowadays. It was that in a deal, the consumer tries to obtain maximum utility, or pleasure, while the entrepreneur seeks maximum profit. Natural as this might look, however, even at the starting point the first serious difficulty appeared. The two centuries since Bernoulli had not been enough to establish what

© Springer-Verlag Berlin Heidelberg 2015
G. Mengov, *Decision Science: A Human-Oriented Perspective*,
Intelligent Systems Reference Library 89, DOI 10.1007/978-3-662-47122-7_2

exactly *utility* was, how it should be defined scientifically, what its properties were, would it be possible to measure it, and how. The need for what today we call an operational definition was evident. In addition, it was necessary to merge theoretically the two kinds of economic agents' motivation—that of a consumer and entrepreneur—because there could not possibly be a fundamental difference between them, and therefore, no justification for more than a single construct existed. Consequently, the two scientists proposed the following definition-like formulation,

> [...] the aim of all participants in the economic system, consumers as well as entrepreneurs, is money, or equivalently a single monetary commodity. This is supposed to be unrestrictedly divisible and substitutable, freely transferable and identical, even in the quantitative sense, with whatever "satisfaction" or "utility" is desired by each participant. (von Neumann and Morgenstern 1944, 2.1.1.)

Neither philosophers nor psychologists, von Neumann and Morgenstern thought it was enough to say that utility is simply what people aim for, with money serving as more or less an intermediary. Over the distance of many decades, we appreciate how they successfully maneuvered around this important conceptual question, yet we are left with the apprehension that utility must be more than just that. In particular, scholars such as Vorobyov (1970) noted that the "monetary commodity" as defined above, certainly does not possess all the properties of money in the real economy. On the other hand, true money does not possess all the characteristics in the above definition either. However, and that is the most important, the commodity in question was sufficiently like money, to be able to serve in that capacity in the new theory, and be identified with money terminologically.

In addition, we should be aware that von Neumann and Morgenstern developed a utility theory intended to serve as auxiliary to their main goal—a *theory of games*, which they planned to make the most adequate description of economic activity in general. Consequently, they introduced the different elements of their definition with the objective to meet the needs of different types of games in the construction they viewed as more important. This is an example how scientists can be careful and selective when preparing the grounds for an ambitious exploratory undertaking.

The new postulate about money and utility being identical was a useful simplification. In their book, the two scholars discussed various alternative ways for defining utility to make it either more plausible or more suitable for their purposes. (von Neumann and Morgenstern 1944, 66.1.1.–67.4.). Vorobyov (1970) suggested in retrospect that they could have introduced an *axiom* about the existence of utility with the above properties, but did not do so due to the controversial nature of the issue. Such an axiom might have become problematic to handle in practical terms, that is, in the mathematical developments of game theory. In addition, the simplification about the equivalence of utility and money, when understood only as a first approximation, avoids asking questions about the exact relationship between the two.

Considerable attention was paid by the two scholars to the question of utility's assessment. At the time, virtually the only certain thing was that if a person had chosen one alternative over another, then for her the first possessed more utility than the second did. An implicit assumption here was that the decision maker is always able to state which alternative they choose. Any measure of usefulness of the utility concept depended on the possibility to make comparisons, which was an accepted view in economics (Samuelson 1938), and one deeply rooted in the method of indifference curves. One unresolved issue at the time was if utility could indeed be measured.

The next assumption was that people are able to express their preferences not only among goods or services, but also among uncertain alternatives. The latter should be understood exactly as defined by us in the opening chapter—as compounds of outcomes with their probabilities. A justification was offered with the argument that uncertainty is inherent in any business activity: the agricultural output depends crucially on the weather conditions; the financial sector is influenced by political events on ongoing basis, etc. Consequently, a reasonable person should be able to make a preference between alternatives such as A and B defined as follows: A: gain of $1000 with probability 0.50 and no gain ($0) with the same probability; B: $450 for sure. Possibly, she may state that she likes them in equal measure. A problem would occur only if, for some reason, the person is unable to make the choice. Another important assumption was that the person is always able to assess the utility of the risky alternative A on its own, in order to be able to compare it with B, the sure thing.

Therefore, this theory linked utilities not with mere goods or services, but with *probabilistic events* occurring in the future and associated with those goods or services. It has been noted (Vorobyov 1970) that here utility is measured with the help of something quite alien to it, as is the probability distribution. Apparently, this approach was influenced in part by the tradition of statistical physics, used by von Neumann as a methodological reference in the development of the new economic theory.

During the late 1940s and early 1950s, this marriage of utilities with probabilities was met with excitement among many economists and decision theorists, because it implied one important consequence—utility was proclaimed measurable. For a long time some of the leading scientists from different generations had been asking themselves if utility was, in their parlance, "ordinal" or "cardinal", i.e. was it qualitative or quantitative. An example of the first would be the beauty of the songs in a rock music top twenty chart—we can neither say exactly how much better the first song is than the second, nor if the distance between the first two is bigger than that between the second and the third. As regards cardinal/quantitative variables, an example would be two opera tickets—we know which one is more expensive and exactly by how much. The question if utility could be measured, i.e. is it cardinal, was given a positive answer by the 19th century theorists of marginal utility. The next generation of economists, however, reversed that position, and then von Neumann and Morgenstern introduced yet another revision, this time armed with a new probabilistic definition and a system of axioms. In retrospect, we know that utility's cardinality and measurability were eventually divorced, and later decades' economics proclaimed utility not measurable once again.

2.2 Axioms and a Theorem in Expected Utility Theory

The axioms of the theory have provoked much interest and have been interpreted and reformulated many times over the decades since their publication. Here, we keep close to their original form. Our objective is to discuss them alongside the way they were subsequently tested by time, as the theory was extensively used in applications and challenged by some of them. It should be noted that the entire system of axioms described alternatives and utilities as perceived by the single person. No comparison of utilities among different individuals was ever implied.

I begin by introducing the following notation. First, p, q, and r are probabilities (real numbers between 0 and 1). Let U be a system of entities denoting abstract utilities u_1, u_2, u_3, ... whose properties are to be clarified. For example, u_1 and u_2 can be the utilities of two mutually exclusive outcomes such as winning \$1000 or winning nothing, as was the content of A above. In U is defined a *relation of preference* ($\succ$) between two utilities, for instance $u_1 \succ u_2$ (that is, u_1 is preferred over u_2. Identical notation is: $u_2 \prec u_1$). In addition, it is assumed that a reasonable person is able to comprehend, and establish for herself a quantitative measure for the total utility of the entire risk-containing alternative. That is equivalent to defining the mathematical operation *combining* of utilities with their respective probabilities: $pu_1 + (1 - p)u_2$, where p and $1 - p$ are the probabilities of the two outcomes. The preference relation and the combining of utilities satisfy the following axioms:

A1. **Complete Ordering**. The relation $u_1 \succ u_2$ is a complete ordering[1] of U. Then

    A1a. For any two u_1 and u_2 one and only one of the following three relations holds: $u_1 = u_2$, $u_1 \succ u_2$, $u_2 \succ u_1$. (Here "=" stands for equal preference, also called indifference.)

    A1b. $u_1 \succ u_2$ and $u_2 \succ u_3$ imply $u_1 \succ u_3$. (Transitivity)

A2. **Ordering and Combining**.

    A2a. $u_1 \prec u_2$ implies that $u_1 \prec pu_1 + (1 - p)u_2$ for every $p \in (0, 1)$.

    A2b. $u_1 \succ u_2$ implies that $u_1 \succ pu_1 + (1 - p)u_2$ for every $p \in (0, 1)$.

    A2c. $u_1 \prec u_3 \prec u_2$ implies the existence of p, such that $pu_1 + (1 - p)u_2 \prec u_3$.

    A2d. $u_1 \succ u_3 \succ u_2$ implies the existence of p, such that $pu_1 + (1 - p)u_2 \succ u_3$.

A3. **Algebra of Combining**.

    A3a. $pu_1 + (1 - p)u_2 = (1 - p)u_2 + pu_1$. (Commutative property)

    A3b. $p(qu_1 + (1 - q)u_2) + (1 - p)u_2 = ru_1 + (1 - r)u_2$ where $r = pq$. (Distributive property)

---

[1]I omit the mathematical treatment of the concept *complete ordering* and appeal to the educated reader's intuitive understanding, for example, about the set of the real numbers—their ordering is complete.

Let me now briefly comment on these axioms and consider some of their implications. It should be noted that in retrospect, it is easy to find fault with them, especially after decades of mounting empirical evidence. All the same, their formulation was a significant step forward in describing economic activity with scientific rigor. Moreover, at least in decision science, though not in economics, the idea that utility can be measured survived the test of time for much longer than the particular axiomatic system.

A1a states that the individual is always able to form an opinion about two alternatives and it can never happen, that she be unable to compare them. If she is unable to choose one, this means that both are equally attractive ($u_1 = u_2$), and by no means that she felt incompetent, or was objectively incompetent to compare them. This is an important limitation of the theory and obviously cannot always be satisfied in practice.

Transitivity, A1b, looks trivial, however it was violated so much and so often, that it became the Achilles' heel of expected utility theory. Indeed, if one prefers A to B and B to C, then one must always choose A over C. Of course, that is unrealistic in many decision situations. A straight example comes from consumer behaviour: goods and services in everyday life cannot be chosen as strictly, because there exists the need for diversity and the desire for changing pleasurable experiences.

Axiom A2a states that if u_2 is preferred over u_1, then each probabilistic combination of the two, which effectively amounts to at least having the chance to obtain u_2, will always be chosen over the less desired u_1. Axiom A2b is dual to A2a.

Axiom A2c essentially states that u_2 may be much desired, but if the chance for it to happen is small enough, its influence on the choice will be neglected. Axiom A2d is the dual of A2c. With these two axioms, von Neumann and Morgenstern in fact introduced a threshold for the perceptions of the agent, who is ignoring too small probabilities, in that case the chance of obtaining u_2. This is remarkable because it effectively states that the decision maker reacts to abstract concepts such as probability in exactly the same way as to sensory stimuli, for which Weber's Law has established the existence of a just noticeable difference (JND). Now, an unexpected implication of the new axioms is that people may be characterized by one more JND—that for too small probabilities. The two theorists never discussed the issue from that perspective in their book. They simply needed this property for further mathematical purposes but arrived, perhaps by chance, at the outskirts of an important psychology-related idea.

In addition, it must be stressed that there is a conceptual difference, but no contradiction, between the postulated human propensity to neglect too small probabilities, and hence, probabilistic outcomes as per A2c and A2d on the one hand, and the property of utility to be "unrestrictedly divisible" according to the theory's defining statement. Utility may be infinitely small, but we can perceive it in finite chunks only.

Axiom A3a states that it does not matter in which order the decision maker would approach u_1 and u_2. The same is true about the more complex combinations of utilities considered in A3b. Both these axioms have also proven problematic in later studies of actual human behaviour. The two authors had a presentiment about such a possibility, as is evident from their suggestion (von Neumann and Morgenstern 3.7.1) that A3b might contradict future empirical findings of a potentially *"much more refined system of psychology [...] than the one now available for the purposes of economics"*. This is yet another example how a pioneering effort can be constrained by the scientific tools of the time.

Using the above axioms, von Neumann and Morgenstern have proven that there exists a mapping of the utilities on the set of real numbers up to a linear transformation. In other words, there exists a mapping between the abstract utility u and the real number v, with the function $v(u)$ having the following properties:

C1. $u_1 \succ u_2$ implies $v(u_1) \succ v(u_2)$.
C2. $v(pu_1 + (1-p)u_2) = pv(u_1) + (1-p)v(u_2)$.
C3. For each two $v_1(u_1)$ and $v_2(u_1)$ holds $v_2(u_1) = a.v_1(u_1) + b$, where a and b are fixed numbers and $a > 0$.

C1 states that the greater of two utilities is ascribed the greater of two numbers, that is, the mapping is monotone. According to C2, the nature of utility is such that the agent is able, loosely speaking, to evaluate the separate utilities of each probabilistic outcome in an alternative, and then combine them to assess the total utility of the alternative. Finally, C3 states that the mapping is determined up to a linear transformation. This means that no absolute zero of the utility is fixed, and no measurement unit is determined. von Neumann and Morgenstern formulated statements C1–C3 as a theorem and provided its detailed proof on about ten pages in an appendix to their book's third edition.

Because we will now accept as a postulate that utility can be numerically represented, for simplicity we abandon the notation $v(u_i)$, which was a shortened version of $v(u(x_i))$, and henceforth will write $u(x_i)$ instead, by which we will understand the utility of x_i. Furthermore, we can perform a natural generalization of C2 to account for not only two, but n possible mutually exclusive outcomes; then the utility of the entire alternative will be denoted as $U(x_1, p_1; \ldots; x_n, p_n)$. This brings us to the definition of utility via Eq. (1.3), which we introduced in Chap. 1 appealing to the reader's intuition instead of resorting to mathematical rigor.

Now we can write down the utility of any alternative, and for example, that of the previously discussed $A \equiv (\$1000, 50\%; \$0, 50\%)$ in the following way:

$$U(A) = \sum_{i=1}^{n} p_i u(x_i)$$
$$= 0.5u(1000) + 0.5u(0).$$

After all this, we can recapitulate what we have done in the following set of statements. If a person chose A over B, then he/she: (1) was able to compare them; (2) for him or her, the first had more utility than the second; (3) the two utilities can be represented numerically; (4) each alternative's utility can be expressed with Eq. (1.3). Formally, all four statements are summarized in this way:

$$A \succ B \Leftrightarrow U(x_{A1}, p_{A1}; \ldots; x_{An}, p_{An}) > U(x_{B1}, p_{B1}; \ldots; x_{Bm}, p_{Bm}).$$

Here, n and m are the number of mutually exclusive outcomes in alternatives A and B respectively. In other words, we have traveled a long way and now are able almost to *compute* utilities of alternatives and then put them in algebraic inequalities to determine the better choice, much as a person is supposedly doing in her or his mind.

Fishburn (1989) has observed that there is a substantial difference between the approach of Daniel Bernoulli, and that of von Neumann and Morgenstern. The former considered utility as subjective for each individual and put it in a probabilistic context, but his analysis mostly emphasized the case of money received, or about to be received, with full certainty. In contrast, expected utility theory as developed by the latter two, deals with probabilistic alternatives, whereby the certain ones are a mere special case. The following example clarifies the issue. Take this equation:

$$u(\$350) = \frac{1}{2}u(\$1000) + \frac{1}{2}u(\$0).$$

In the von Neumann–Morgenstern interpretation the above means that a person considers equally attractive the sure \$350, on the one hand, and on the other, the risky choice of \$1000, to be gained with probability 0.50, or \$0 with the same probability. Now look at the equivalent algebraic equation:

$$u(\$350) - u(\$0) = u(\$1000) - u(\$350).$$

Its Bernoullian interpretation involves no probabilities but only sure gains. The agent derives utility from the first \$350 that is exactly equal to the utility of the subsequently received \$650, if she already had received the first \$350. Here we used the additional assumption that $u(\$0) = 0$.

2.3 Evolution of the Theory: Allais' Paradox

In the 1940s, many readers of Theory of Games and Economic Behaviour were impressed by the new horizons for exploration this volume outlined. Among other issues, one old question resurfaced: Now that utility was cardinal again, could it be possible to invent a method for measuring it, and then quantify the entire utility function of a decision maker? An important objective was to develop a procedure that would link the agent's revealed preferences—an established concept in economics due to Samuelson (1938)—with the probabilistic alternatives now at the centre of the attention. The matter had huge practical consequences. In particular, some economists were eager to use the new theory to understand why people are willing to buy insurance policies, i.e. to pay to eliminate certain risks, and at the same time engage in activities such as betting on sporting events, buying lottery tickets, and investing in stock shares—all of which amount to gambling, i.e., buying risks.

An article by Friedman and Savage (1948) from the University of Chicago made a pioneering contribution to the understanding of those issues. The two proposed a method eliciting the decision maker's preferences and leading to the construction of his or her utility function. They started with an example such as this: You can have an income of $500 or $1000 with probabilities p and $(1 - p)$ respectively. What value of p will make that alternative as attractive as a fixed income of $600? Alternatively, one might fix the probability, e.g., $p = 0.30$, and then ask what would be the certainty equivalent of the risky alternative, in that case ($500, 30 %; $1000, 70 %)? Having obtained the answers to many such questions from a person, the analyst would be able to construct an empirical curve of that person's utility function. Friedman and Savage went further to suggest that having such knowledge about the preferences would make it possible to predict individual reactions in any such future situation involving risk. Remarkably, that idea remains at the core of all methods for eliciting personal utility curves even in the 21st century.

Attempting to explain why people are willing to spend money to avoid risk, and at the same time to acquire some forms of risk, the two scholars hypothesized a certain shape of the utility curve that would account for such behaviours. They suggested that utility as a mathematical function of income must consist of three segments as shown in Fig. 2.1. First, there is an initial region where the function is concave upward, which corresponds to risk avoiding.[2] That can be intuitively understood by recalling the St. Petersburg Paradox (Box 1.1): because of the diminishing marginal utility of money, people would prefer to sell their right to gamble and get a smaller but certain sum instead of a potentially huge but uncertain one. In this vein, a low-income agent would be willing to avoid risk and buy insurance, provided it is not too expensive. The second segment is convex, indicating risk seeking behaviour that in practical terms amounts to paying for the opportunity to gamble. Finally, in the third region, the costs of risky behaviour begin to look too great and gambling is again mostly avoided.

Friedman and Savage even speculated that the middle segment might relate to buying lotteries, offering a chance for a huge gain that would lift the lucky person out of their social class and put them in an upper class. Such behaviour is plausible, they mused, because, *"Men will and do take great risks to distinguish themselves [...]"*. In retrospect, we know that neither that particular suggestion, nor the

---

[2]Here I omit the mathematical proof of the equivalence of risk avoiding and concavity of the utility function. Interested readers are referred to Eeckhoudt and Gollier (1995) or Mas-Colell et al. (1995).

Fig. 2.1 The agent's utility
curve according to Friedman
and Savage (1948)

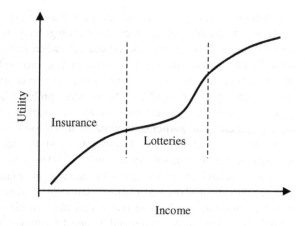

three-segment curve survived the test of reality, but this scientific field was just taking off, and it was the time of charming new ideas.

Further, in the middle of the 20th century it looked implausible that people had an apparatus in their brains to calculate risky utilities as a function of their income. In addition, no one could imagine a technology to look into one's brain and verify such a hypothesis. On the other hand, observing someone who always prefers A to B, B to C etc. regardless of how many times she has to choose, may lead us to believe that she has some brain mechanism doing utility maximization, which could be perfectly described by some mathematical function.

With such ideas in the air, it was certainly tempting to devise a real experiment aimed at measuring utility. Frederick Mosteller and Philip Nogee of Harvard University did exactly that (Mosteller and Nogee 1951). Their work went down in the history of decision science perhaps less recognized than it deserved, but it certainly contained a number of accomplishments. Probably the most important was the first empirical assessment of a personal utility function, as we understand it today. Two groups of subjects participated in their experiment—Harvard under-graduates and military professionals. Personal as well as between-group differences were hypothesized due to the participants' different social profiles: the students were younger, financially more privileged, and more optimistic about their future income. It turned out that these differences were indeed influencing the utility curves, but no clear profiles could be established.

Secondly, this was one of the earliest experiments ever, in which participants were paid in proportion to the results of their decisions, rather than for the time they spent in the laboratory, as was the norm for psychological experiments. In that sense, this was one of the earliest economic experiments conducted in history. It is interesting that even today economists remain skeptical about the quality of experimental work done by psychologists on the sole ground that money is not used in the proper way to motivate people. Of course, psychologists have their deep reasons to dismiss such an attitude.

Mosteller and Nogee's third achievement was the experimental confirmation of a result already established by other scientists regarding the way people comprehend probabilities. It had recently turned out that when choosing among risky alternatives, the human mind does not use mathematical probabilities straight as given, but "twists" them, that is, changes their values and eventually takes decisions using these altered "psychological", or "subjective" probabilities.

With regard to expected utility theory, very significant was the fourth result of the experiment. The participants violated much too often some of the axioms, in particular A2 (Ordering and Combining) and the requirement for stability of preferences, i.e. the transitivity axiom (A1b). Of course, no one expected totally consistent behaviour when choosing among alternatives. von Neumann and Morgenstern themselves had claimed that their system of axioms was just satisfactory. Friedman and Savage had called the transitivity requirement "an idealization". However, the Mosteller and Nogee data showed that expected utility theory would not be able, in its current form, to accommodate many of the newly established facts about real economic behaviour. In addition, that first experiment offered some support for, but at the same time cast some doubt on the Friedman–Savage three-segment utility curve.

The theory now shaken, it encouraged other scientists to question its validity on other accounts. The French economist Maurice Allais dealt it a substantial blow by showing that people violate the transitivity axiom A1b in a *systematic* way. His discovery remained in history as *Allais' Paradox*, which may be summarized as follows. Let a person have to choose between the following two alternatives:

$$A1 \equiv (\$1\,\text{m}; 100\,\%) \quad \text{and} \quad B1 \equiv \begin{pmatrix} \$1\,\text{m}; & 89\,\%; \\ \$5\,\text{m}; & 10\,\%; \\ \$0; & 1\,\% \end{pmatrix}.$$

Most often, he/she chooses the sure one million of $A1$. However, when the same person has to choose between these alternatives:

$$A2 \equiv \begin{pmatrix} \$1\,\text{m}; & 11\,\% \\ \$0; & 89\,\% \end{pmatrix} \quad \text{and} \quad B2 \equiv \begin{pmatrix} \$5\,\text{m}; & 10\,\% \\ \$0; & 90\,\% \end{pmatrix}$$

$B2$ is almost always the preferred option. However, the alternatives in the second problem are simply two dented versions of those in the first problem: indeed, to obtain $A2$ and $B2$ one must remove the opportunity to win \$1 m with probability 89 % from both $A1$ and $B1$. Taking away the same amount of utility from two competing alternatives should not change the decision maker's preferences between them. Let us use the main theorem (C1 and C2) as well as Eq. (1.3), with the natural assumption that zero gain has zero utility $u(0) = 0$, and also substitute $x_1 = \$1\,\text{m}, x_2 = \$5\,\text{m}, x_3 = \$0$, $p_1 = 89\,\%, p_2 = 10\,\%, p_3 = 1\,\%$. Then the preference of the majority of people, $A1 \succ B1$, can be expressed as an algebraic inequality:

$$p_1u(x_1) + (1 - p_1)u(x_1) > p_1u(x_1) + p_2u(x_2). \tag{2.1}$$

Similarly, from $A2 \prec B2$ we obtain:

$$(1 - p_1)u(x_1) < p_2u(x_2). \tag{2.2}$$

Since it is not possible to reverse the sign of Eq. (2.1) by subtracting from both sides the positive quantity $p_1u(x_1)$, it follows that Eqs. (2.1) and (2.2) cannot be simultaneously true. This implies that in the second problem people must prefer $A2$ to $B2$. In reality, an overwhelming majority chooses the opposite, and there lies the paradox.

Allais has demonstrated another kind of preference reversal, presented here with a slightly different set of alternatives than the original. When choosing between

$$A3 \equiv (\$3000; 100\%) \quad \text{and} \quad B3 \equiv \begin{pmatrix} \$4000; & 80\% \\ \$0; & 20\% \end{pmatrix},$$

people usually opt for $A3$. A fourfold reduction of the odds, in the first alternative from 100 % to 25 %, and in the second from 80 % to 20 %, would lead to the following new alternatives: $A4 \equiv (\$3000; 25\ \%)$ and $B4 \equiv (\$4000; 20\ \%)$. Instead of retaining their preferences, people again reverse them and en masse opt for $B4$.

Allais compiled a list of such problems and disseminated it among his students and colleagues. The majority of them fell in the traps and, by and large, violated the von Neumann–Morgenstern axioms. During a lunch at a 1952 scientific conference in Paris, Allais offered the problems to Leonard Savage himself, who also gave contradictory answers, much to his own amazement. After the initial shock, he decided to change some preferences to stay consistent with his own theory, rather than fall prey to Allais' paradox. Apparently problematic as a description of actual economic behaviour, expected utility theory, he believed, was still adequate as a normative theory (Shafer 1984).

2.4 New Issues

Since the discovery of Allais' paradox, the theory was destined to develop amid clashes of its proponents and its critics which continue to this day, although with diminished intensity. The debate was still in full force when in 1979 two important publications attracted attention. The first (Allais and Hagen 1979) was a collection of articles comprising a kind of recapitulation of the achievements and accumulated issues over the preceding three decades. In it, Allais gave an exposition of his own theory of decision-making under risk, while Morgenstern defended expected utility theory claiming that it was adequate descriptively, had a limited domain of applicability where it was accurate, and in that sense could even be compared to Newtonian mechanics. That volume contained also some empirical research by other contributors hinting at the psychological invasion that was to happen in the domain of decision science.

The second publication appeared in the leading journal *Econometrica* and presented *Prospect theory: an analysis of decision under risk* by psychologists Daniel Kahneman and Amos Tversky. By 2002, the year when Kahneman was awarded the Nobel Prize, this was the most cited article in that journal. It literally overhauled the scientific understanding of the way people take decisions, and remains highly influential to this day. I leave its detailed discussion to Chap. 3, and now continue with the developments around expected utility theory.

Its proponents adopted two different strategies to defend it, not counting the third one—of simply ignoring all criticisms. The first strategy was to formulate new axioms weakening the theory's strict requirements and reconnecting it with the observed economic behaviour. The second strategy was inspired by Savage's reaction to Allais' paradox (he revised his preferences to remain faithful to his principles). Born spontaneously, this tactic was later recognized as a potentially powerful method to achieve optimality in decision-making. This is how its followers reason: Indeed, one cannot always analyze the options available with utmost precision, and does not always make the best choice. However, such inefficiencies are caused by occasional inattentiveness, are accidental, and can be corrected. To this end, a more careful second thought would virtually always be sufficient.

Adding to these arguments from a recent perspective, we can say that in our time the agent can receive assistance by a computer-based decision support system, a consulting expert system, or something more modern in the same vein. Generally, the resulting procedure would involve reexamining the answers a person gave and finding out the contradictions among them. Then the latter—presumably a small number—will be revised to achieve a kind of "global consistency".

Of course, this will not be relevant to the thousands of minor decisions of little consequence we take on daily basis. Some of them can contradict our general pattern of activity but we may never become aware of that fact. Fortunately, almost all of our actions will be satisfactory, and even close enough to optimality due to inbuilt cognitive mechanisms.

The use of computer-aided consulting, however, requires a substantial effort in terms of time and financial means. Doing it makes sense only for important problems that are difficult to solve. Note that this is not the case when we seek professional advice in areas outside of our expertise—for example, when we need medical help, or assistance with new technology or software. Rather, the situation is much as we are the expert, but need to take into account a large variety of circumstances and have to use a computer model to help us. We can clarify this issue by adopting for a moment the point of view of the developers of such expert systems. For instance, an experienced physician might be asked to provide assessment on a multitude of medical cases because his knowledge will be put in a software to train students. However, the algorithm has discovered some inconsistencies among his answers and he has to reexamine them and introduce the needed corrections. After all, the value of the system will be in providing help with the really intricate cases.

This approach is most convincing when decision analysis tries to help solve important societal problems. When the optimal solution depends on a large number of criteria, the superiority of one alternative over the remaining is not obvious and may often be counterintuitive. In such cases it is said that decisions are not taken, but are computed. Pioneering research in that area was conducted in the 1960s at Harvard and MIT, with its high point being a book by Keeney and Raiffa (1976). Its topic was how to take decisions with multiple conflicting objectives especially when the latter are incommensurable. This may involve defining axiomatic systems to guide preferences among complex risk-containing alternatives. Then the researcher has to do theoretical work to find suitable special cases of otherwise intractable problems involving multidimensional utility functions. Finally, methods and procedures must be developed (Fishburn 1977) to extract those utility functions from the social values of the decision makers. The field was hot in the late 1970s as another book of the same school (Bell et al. 1977) summarized a number of subsequent theoretical contributions and practical cases. At the time, an already impressive list of applications existed. Such methods had been used in decisions involving: the construction of Mexico City Airport; selection of sites for nuclear reactors in the State of Washington; urban planning for the city of Darmstadt, Germany; resolving ecological issues around Japanese power plants; forestry management in New Brunswick, Canada; and many others. All these cases involved direct applications of expected utility theory and its extensions, and to this day serve as primary examples for rationally taken important decisions.

Now let us come back to the other strategy adopted by some theorists: weakening the original axioms to accommodate the most prominent violations of the theory. A huge step was made when the mathematical probabilities were replaced by "subjective" or "psychological" probabilities in the decision models. That idea was not new. As early as 1926, the Cambridge mathematician Frank Ramsey invented the concept of subjective probability, formulated axioms about it, and even suggested a way to measure it. Unfortunately, Ramsey died at the age of 26 and his ideas went into obscurity, even escaping the attention of von Neumann and Morgenstern (Fishburn 1989).

However, towards the end of the 1940s a lot of empirical evidence showed that people indeed distort numerical probabilities when thinking about them. Perhaps the first to discover this experimentally were Preston and Baratta (1948), who in 1948 observed and documented how gamblers in a lab game treated a mathematical probability of 0.75 as if it were equal to 0.61. Discrepancies existed also for other values, with people estimating objective probability reasonably adequately only around $p = 0.20$. In another study, Griffith investigated data from horseracing betting and established a corresponding point of equivalence at around 0.16. The Mosteller–Nogee experiment did not confirm any of these findings, but suggested an interval of 0.10–0.55 in which such an equivalence point might exist. All this was enough for Allais (1953) to use it as another argument against expected utility theory.

Naturally, the issue came across the attention of psychologists. One of them, Ward Edwards, the scientific mentor of Amos Tversky, discovered that in betting people prefer to deal with some probability values more often than with others. For example, they consistently liked bets with probability 0.50 but tended to avoid those with 0.75. Those preferences were reversed when losses were involved. Edwards suggested that not probability, but some extension of it, a kind of *weighted probability* should be incorporated in the analytic models.

Savage (1954) formulated a *subjective expected utility theory* for decision-making under uncertainty, in which he developed a system of axioms and explicitly used subjective probability. In the 1950s and 1960s, a new wave of mathematical models considered subjective probability $\pi(p)$ as a function of the mathematical probability p, whereby the empirical curve was built using a procedure eliciting a person's preferences. In a 1989 survey, Machina (1989) put together and compared at least five such models and their variations.

How objective and subjective probability are related, is clarified in sufficient measure in Kahneman and Tversky's prospect theory (1979, 1992). They needed that particular knowledge to define the general utility of an alternative (or prospect) with a new and more complex model, which was this:

$$U(x_1, p_1; \ldots; x_n, p_n) = \pi(p_1)u(x_1) + \ldots + \pi(p_n)u(x_n). \tag{2.3}$$

Comparing Eq. (2.3) with Eqs. (1.2) and (1.3) shows how the former introduces a further weakening of the mathematical expectation principle. What were once straight monetary gains and ordinary probabilities, were later diluted with Bernoulli's subjective utility of the gains, to be further complicated with subjective probabilities. This is a classical scientific approach—where a simple model cannot explain a phenomenon, a more sophisticated one is invented to do the job. As we will see, the innovation with introducing $\pi(p)$ helped resolve Allais' paradox, but opened the door for new paradoxes.

2.5 Empirical Assessment of Utility

For many decades, scientists have had a rough time trying to understand human choices not least because of the inherent problems with the idea for utility maximization. The fact remains, that we still do not have a sufficiently good understanding of what utility u is. It certainly helped that von Neumann and Morgenstern simplified matters by amalgamating money (the "monetary commodity") with utility, and in this way achieved a "satisfactory" and maybe even impressive account of many forms of economic behaviour. In fact, when the economic agent aims at receiving a substance both equivalent to money and bringing satisfaction, this is a very good combination of a theoretical and operational definition. Exactly that was what opened the way for experimental work and further advancement of not one, but two theories—of games, and of expected utility.

Today, however, a lot of new knowledge about humans as decision makers has accumulated and the vagueness of the utility construct is no longer acceptable. For pragmatic reasons, many scientists have adapted to its being measurable up to a linear transformation only. This circumvents the assessment problem and opens the door for the revealed preferences approach, avoiding the issue of utility's content. Indeed, the Friedman–Savage idea to interview a person about their preferences among probabilistic alternatives can be very constructive. A cleverly selected set of monetary gains, losses, and their probabilities can do a lot to uncover interesting facts about a person's utility function and even something about the motivation behind it.

In particular, changes in the curve's steepness may be quite illuminating. If it is natural to associate the curve's concavity with diminished marginal utility of money, and with risk-avoiding behaviour, then how a local curve convexity must be comprehended? In 18th century, Bernoulli himself gave an example with somebody jailed for unpaid debts. He suggested that a wealthy prisoner who needed two thousand ducats in addition to what he possessed to repurchase his freedom would derive more utility from that sum than somebody much poorer would. Under normal circumstances it will be, of course, the other way round—that money would bring more utility to the poor person. Graphically, this effect is a locally convex segment of the utility curve when approaching 2000 ducats, followed by the normal concave continuation.

Box 2.1 Individual utility curve with a local convex anomaly
Today we possess methods to obtain individual utility curves. In 2005, my colleague Yuri Pavlov from the Bulgarian Academy of Sciences and I conducted a computer-based experiment to characterize the utility curve of a German Professor of Economics. We used an iterative stochastic algorithm, developed by Dr. Pavlov (1989, 2005).

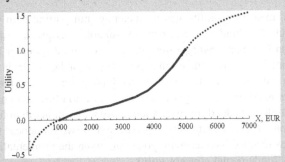

The experiment took place after the professor lectured undergraduates on the concepts of utility, marginal utility, and the First Law of Gossen. Then he engaged in our computer-based procedure and soon the result was available. Unexpectedly, the curve that emerged was quite different from what the theory would predict. Indeed, in the segment €1000–2500 it was concave (see the figure), but immediately afterwards it rose steeply and became convex up to €5000. Possibly, it could have continued that way, had we not stopped the procedure, considering inappropriate to subject the professor to further experimenting. Neither did we ask about the potential acquisition he had in mind. Economists, who saw that curve, were puzzled by its anomalous form and suggested that we draw two hypothetical concave segments with dotted lines around it, to make it look more in line with the established views.

The fictitious units on the y-axis of the figure in Box 2.1 remind us how problematic the measurement of utility is without a convincing operational definition. An influential study by Alchian (1953) produced the puzzling result that, proclaiming utility *measurable* did not help the economic analysis at all and was, moreover, methodologically unjustified. His main objection was that there existed no method for assessing the utility of a good in a market basket in conjunction with the utilities of the remaining goods in the basket. It was impossible to sum them up to obtain the total basket utility. (Will eating more chocolates, or meatballs, or both, bring to somebody more utility?) In addition, no prediction of somebody's future choice based on assessment of the alternatives' utilities was feasible.

The solution that economists found to this long-standing issue was to divorce cardinality from measurability. Alchian (1953) was quite clear about the latter, but it would have been too detrimental to discard the von Neumann–Morgenstern approach altogether, not least because of the benefit of the mathematical tools that came along with it. Gradually and quietly, economics decided that,

> Cardinality of a utility function is simply the mathematical property of uniqueness up to a linear transformation. [...] cardinality and measurability [...] were different concepts, and the former in no way implied the latter. (Lewin 1996)

In other words, measuring utility has no meaning, but putting it in mathematical models is useful. The kind that was actually suitable was given a specific name: *vNM utility* (von Neumann–Morgenstern utility). A neutral observer may find this slightly odd, but has to remember that progress must be made in one way or another, and in the social sciences such developments seem to be the norm rather than the exception. After all, as regards economics, Milton Friedman had concluded that a theory should be judged for its predictive capability rather than for the realism of its assumptions. From a 21st century perspective, it would appear that expected utility theory has reached its maximum potential, given the problematic operational definition of its fundamental concept *utility*.

2.6 Arrow–Pratt Formula for the Price of Risk

One of the most important achievements of utility theory was the quantification of the price a person is willing to pay to buy certain amount of risk, or alternatively, to get rid of a risk. In practical terms, such behaviour happens always when somebody indulges in the pleasure of gambling, or prefers to stay too much on the safe side.

Let a financier's job be to choose many times every day between $A \equiv$ (€1000, 50 %; €0, 50 %) and B: a sure gain of €450. If he consistently chooses A, at the end of a long period, say a year, he will have from each deal an average gain of around €500. Should he play it safe, he will have in each case €450. The question is more complicated if the decision maker faces the choice not on daily basis, but just once—then he might find it quite hard to decide. In both cases, however, we say that if he chooses A, he is probably *risk-neutral* while choosing B defines him as *risk-averse*, or a *riskophobe*. Anybody who practices some form of gambling, which technically means to pay more than the mathematical expectation of the risk-containing alternative, are termed *risk seeking*, or *riskophile*.

The price we are willing to pay to acquire a risk, or to get rid of a risk, is called *risk premium*. Independently of one another, in 1964, John Pratt (1964) and Kenneth Arrow (1965) came at almost identical results about the analytical definition of this quantity. Here we derive the Arrow–Pratt formula, following in part Eeckhoudt and Gollier (1995).

We use the following notation. Let W_0 be the initial wealth of a person who has to make a choice about a risk-containing alternative A, which could be the one above, but could be any alternative of the kind $A \equiv (x_1, p_1; \ldots; x_n, p_n)$. In fact, A can be not only a discrete variable, but a continuous one with a probability density function $f(x)$. Let it be defined in the interval $x \in [a, b]$ where a and b are the minimum and maximum possible gains (or losses). With no loss of generality, but for better intuitive understanding, we assume that the mathematical expectation of the alternative is positive.

Suppose that a person's wealth consists of two parts: one certain, W_0, as described already; and one risk containing, A, which may be shares, traded in a stock exchange. Then her total (final) wealth W_f is:

$$W_f = W_0 + A.$$

The mathematical expectation of A in the discrete case will be:

$$\mu \equiv E(A) = \sum_{i=1}^{n} p_i x_i, \tag{2.4}$$

while in the continuous case, given the domain $[a, b]$ in which it is defined, it is:

$$\mu \equiv E(A) = \int_{a}^{b} x f(x) dx. \tag{2.5}$$

Then the mathematical expectation of the total wealth W_f will be

$$E(W_f) = E(W_0 + A) = W_0 + E(A).$$

The person derives from that wealth W_f utility U, which can be described in the discrete case as

$$U(W_f) = \sum_{i=1}^{n} p_i u(W_0 + x_i)$$

and in the continuous case, as

$$U(W_f) = \int_a^b u(W_0 + x)f(x)dx. \tag{2.6}$$

For mathematical correctness, here we adopt the following notation. As usual, U denotes the utility of the entire risk-containing alternative; u is the utility of a single outcome, or the utility of the entire alternative when used in the integrand of equations with U in the *lhs*. The latter case refers only to Eqs. (2.6), (2.7), and (2.10).

Now suppose that the person is uncomfortable about the risk contained in the stock shares and would like to sell them. Naturally, after the transaction she must derive the same utility from her total wealth as before it. Let us denote this risk-free equivalent of her wealth as W^*. Because the total utility must not change, it is necessary that $u(W^*) = U(W_f)$. We substitute this in Eq. (2.6) and obtain:

$$U(W^*) = \int_a^b u(W_0 + x)f(x)dx. \tag{2.7}$$

The person will exchange the risk-containing wealth $(W_0 + A)$ for the risk-free wealth W^* at a fair price. From her point of view, the price for A can be defined as:

$$P_a = W^* - W_0. \tag{2.8}$$

In fact, the "fair price" P_a fixes the minimum amount of money the person would be willing to accept for A to make the deal. A theorem (omitted here) states that if the person is risk neutral, her asking price P_a will be equal to the mathematical expectation of the risk-containing alternative A. Therefore, for her holds $P_a = E(A)$. If the person is a riskophobe (which she is in the example), or a riskophile, $P_a \neq E(A)$ will hold.

Now we are ready to define *risk premium* η with the following equation:

$$\eta = \mu - P_a, \tag{2.9}$$

where μ is defined by Eqs. (2.4) and (2.5). Apparently, the risk premium shows how much the person is willing to give out just to get rid of the risky stock. In our illustrative case $\eta > 0$, which means that $P_a < \mu$ or in other words, the person will accept a sum of money smaller than μ. To reiterate, because the person is a riskophobe, she is willing to get rid of the risky stock A and to this end will accept a fair price P_a, which is *less* than the objective mathematical expectation of A. Hypothetically, the person could be neutral to the risk and choose to ask for price exactly equal to the mathematical expectation, rendering the risk premium equal to zero. However, we are interested in quantifying η in the general case. To this end, we rewrite Eq. (2.8) as

$$W^* = W_0 + P_a$$

and with its *rhs* we substitute W^* in the utility function's argument in the *lhs* of Eq. (2.7). We obtain

$$U(W_0 + P_a) = \int_a^b u(W_0 + x)f(x)dx. \tag{2.10}$$

In general, regardless of her being risk neutral, riskophile, or riskophobe, if she received an offer below her P_a, she would reject it, lest her total utility diminished. Our objective is to determine that price P_a from Eq. (2.10). We use the fact that each mathematical function of certain properties (being n-times continuously differentiable etc.) can be approximated with a Taylor series. We recall that $\mu \equiv E(A)$ and expand the *lhs* of Eq. (2.10) around the point $(W_0 + \mu)$. Note that this particular point is in a sense the most objective assessment of the person's total wealth. A first-order approximation will give:

$$\begin{aligned} U(W_0 + P_a) &\approx U(W_0 + \mu) + (W_0 + P_a - (W_0 + \mu))U'(W_0 + \mu) \\ &\approx U(W_0 + \mu) + (P_a - \mu)U'(W_0 + \mu). \end{aligned} \tag{2.11}$$

Similarly, we expand $u(W_0 + x)$ from the integrand in Eq. (2.10), now reverting to its original notation $U(W_0 + x)$, around the point $(W_0 + \mu)$. Because $(W_0 + A)$ can vary around $(W_0 + \mu)$ a lot more than $(W_0 + P_a)$ can, now a higher-order approximation will be needed. That is why we will take the first three terms in the series:

$$U(W_0 + x) \approx U(W_0 + \mu) + (x - \mu)U'(W_0 + \mu) + \frac{(x - \mu)^2}{2!}U''(W_0 + \mu). \tag{2.12}$$

We have to substitute the integrand of Eq. (2.10) with the *rhs* expression of Eq. (2.12). In doing so we keep in mind that because of the definition of μ by Eq. (2.5), the following holds:

$$\int_a^b (x - \mu)f(x)dx = 0,$$

and also, we use the variance definition

$$\int_a^b (x - \mu)^2 f(x) dx = \sigma^2.$$

After the necessary transformations, for the *rhs* of Eq. (2.10) we obtain:

$$U(W_0 + \mu) + \frac{\sigma^2}{2} U''(W_0 + \mu). \tag{2.13}$$

Now, we substitute the *lhs* and *rhs* of Eq. (2.10) with the expressions from Eqs. (2.11) and (2.13) respectively, to obtain, after short transformations:

$$P_a - \mu \approx \frac{\sigma^2}{2} \frac{U''(W_0 + \mu)}{U'(W_0 + \mu)}. \tag{2.14}$$

Remembering the defining Eq. (2.9) for risk premium, and using Eq. (2.14), we finally obtain

$$\eta \approx \sigma^2 \left[-\frac{1}{2} \frac{U''(W_0 + \mu)}{U'(W_0 + \mu)} \right]. \tag{2.15}$$

Equation (2.15) is the Arrow–Pratt approximate formula for the risk premium. Let us discuss it briefly. First, the risk premium is proportional to σ^2, which may be considered as a measure of financial uncertainty, although a crude and rudimentary one. The proportionality means that the greater the risk, the higher the premium a risk-averse person will be willing to pay to get rid of it. In the same way, a riskophile will be willing to pay a *higher* premium for a *greater* risk, attracted either by the greater potential gain or simply by the prospect for more pleasure in gambling.

In Eqs. (2.14) and (2.15), the quantity

$$-U''(W_0 + \mu)/U'(W_0 + \mu)$$

is called *degree of absolute risk aversion*. It was introduced and initially studied in 1961 by Robert Schlaifer from Harvard University (Pratt 1995). It depends on U (W_0) and is therefore unique for every person. Two people possessing the same wealth W_0 and finding themselves in the same risk-containing situation, will take different decisions (will pay different risk premium) because of their different utility functions. That quantity is also a *local* measure of risk aversion in the sense that it in no way characterizes the entire utility function.

Further, because Eq. (2.15) was derived using approximations it is applicable to cases where the outcomes in the risk-containing alternative are significantly smaller than the person's total wealth. The greater the risk, the greater the formula's error, with one exception, which is the risk neutral individual for whom the formula is always accurate and gives risk premium equal to zero.

Example Let us illustrate the use of the Arrow–Pratt formula with a numerical example. Imagine that the financier from the beginning of this section is considering alternative A $\equiv$ (€1000, 50 %; €0, 50 %). Let his utility function in the interval [€0, €1,000,000] be given by the formula:

$$U(W_0) = W_0 + a \sin(W_0/a), \qquad (2.16)$$

where a is a real positive constant. For $a = 10^5$, the financier's utility curve is shown in Fig. 2.2. Obviously, that kind of shape can be defined in infinitely many ways different from Eq. (2.16). Here we have a utility function with total wealth as argument, unlike the Friedman–Savage idea in Fig. 2.1, which plotted utility vs. income. I avoid the economics debate regarding which—income or wealth—should be the argument in the utility function, because it is irrelevant to the example with the particular alternative A. Psychologically, the difference for the decision maker cannot be very significant due to the very small amounts of money in A.

Now, if the financier has wealth W_0 amounting to €100,000 as shown in Fig. 2.2 by the first dotted line to the left, his utility function is concave there, and so he is expected to be risk averter ready to pay a premium to get rid of alternative A. Around €300,000 he may be even more inclined to avoid the risk (the dotted line in the middle), while for $W_0 =$ €500,000 some appetite for risk may develop, as the convex curve intersected by the right dotted line suggests. In that case he may be willing to pay a premium when buying A.

Let us check these three suggestions applying the Arrow–Pratt formula. First, we have to calculate μ and σ^2 for the prospect A. We use Eq. (2.4) to obtain $\mu = 500$, which then goes in $\sigma^2 = \sum_i (x_i - \mu)^2 p_i$ to compute $\sigma^2 = 250,000$. Then we differentiate twice the utility function from Eq. (2.16):

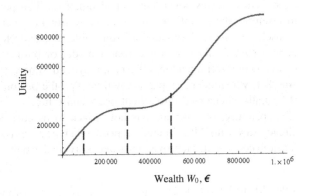

Fig. 2.2 A hypothetical utility curve of a financier

$$U'(W_0) = 1 + \cos(W_0/a),$$
$$U''(W_0) = -(1/a)\sin(W_0/a).$$

With these derivatives we substitute in Eq. (2.15) the respective quantities. We compute the risk premium values for the particular a, μ, σ^2 and the three discussed cases of W_0. The results are shown in Table 2.1.

The figures in Table 2.1 can be interpreted as follows. If the financier had a sure wealth of €100,000 and in addition, possessed A, he would have been ready to sell it for €499.31 (because $\mu = 500$). Were he three times richer, he would have been even more risk averse, willing to sell it for mere €481.73. However, with a personal wealth of half a million, but no possession of A, he would have offered €500.93 to buy it. He is a bit of a gambler now.

The Arrow–Pratt approach became widely used in a number of areas of economic theory. Only about a decade after it was published, it was already present in studies of demand in insurance and asset markets, taxation models, savings models (Ross 1981) and many other domains. Where it was found problematic, it was revised and extended, and also used as a reference point for similar approaches. There have been attempts even to estimate empirically the Arrow–Pratt coefficient of absolute risk aversion for various kinds of economic agents, including agricultural producers in some States in the US (Wilson and Eidman 1983). Hypotheses about the relative risk aversion (equal to the absolute, multiplied by the total wealth) of the representative investor in the US securities market have been empirically tested (Dieffenbach 1975).

Let us ask the question to what extent this type of analysis reflects people's actual behaviour. Even the most assertive economic applications have left the door open, admitting that empirical proofs have never been wholly satisfactory. As usual, psychologists had something to contribute to the understanding of the issue. It might be appropriate to discuss their point of view first by looking at the mere labels "riskophile", "riskophobe", "risk-neutral individual". Implicitly, they all suggest that people have a stable feature with regard to risk bearing, which is expressed somehow consistently in all situations. This personal characteristic ought to resemble the established in psychology Big Five personality traits: Openness, Conscientiousness, Extraversion, Agreeableness, and Neuroticism—they all comprise a statistically discovered robust model of human personality. While all of them change over a person's lifespan, they tend to be very stable during adulthood. Another, yet simpler comparison will be with the person's color of the eyes which is virtually unchanged throughout one's entire life.

Apparently, risk attitude has not proven to be such a stable construct. As we already saw with Allais' paradox, people's choices in risk-containing situations are far from consistent. Moreover, the same individual usually shows different degrees of

Table 2.1 Risk premium η for prospect A at three different levels of wealth

W_0	€100,000	€300,000	€500,000
η	€0.69	€18.27	−€0.93

risky behaviour in various aspects of their life—personal, professional, social etc. Therefore, trying to anticipate reactions to risk, one has to use measurement tools with tasks as similar as possible to the actual cases of interest. A number of studies have shown how specific and intricate that can be. For example, Weber et al. (2002) showed that real-world investment decisions were predicted well by psychometric instruments involving also investment decisions, but were predicted poorly when gambling tasks were used, although both experimental treatments involved monetary outcomes. In short, people's mindset regarding risk is highly context-dependent.

2.7 Experienced Utility and Future Utility

A further example of that context sensitivity is the distinction between two types of utility. A postulate in the von Neumann–Morgenstern theory was that the decision maker chooses among probabilistic alternatives with potential gains and losses, and expects the outcome in the near future (von Neumann and Morgenstern 1944). Quite a useful simplification, undoubtedly. It turned out, however, that people assess in entirely different ways the usefulness of past decisions as opposed to imminent prospects. Two very different attitudes play a role here. Past decisions have led to certain experiences, associated with pleasure, pain, and emotions. This is experienced utility. In contrast, present choices are governed by anticipation regarding future gains. Expected utility theory implicitly considers both types of utility being identical. However, this might be the case only for somebody with the gift of predicting precisely what pleasure, benefit or another type of advantage will the chosen prospect bring. But does such an ideal hedonist exist? The von Neumann–Morgenstern theory assumes that the answer is yes, although the two scholars never discussed it explicitly and there is no reason to believe that they ever gave it a thought. It looks more as we have here what Poincaré called a *hidden axiom*, one that is operating without the theoreticians being aware of its existence.

The problem with the ideal hedonist was noticed by James March (1978) and later elaborated by Tversky and Kahneman (1981) and Kahneman and Tversky (1984). Common sense tells us that such a character is impossible. A popular and vivid example (Kahneman and Tversky 1984) is the hungry man who has ordered five meals, only to discover how wrong he had been by the time the fifth one is served. It is indeed very hard to predict what exactly a choice will bring to us. The problem is even more complicated due to the importance of the context and wording used to describe the needs and the options available to satisfy them. Tversky and Kahneman (1981) suggested a pragmatic advice to deal with such situations: instead of asking "What do I want now?", it would be more useful to adopt a prognostic orientation and ask "What will I feel then?" Hopefully, with more experience the answers to the second question will tend to become more accurate.

In fact, as early as the 17th century, Blaise Pascal noticed the existence of the two types of utility in a rudimentary form as he observed how gamblers derive pleasure from the game independent of their actual gains. Sometimes other examples in this vein are given, like the mountain climbers who relish a dangerous experience bringing no apparent benefit. From a narrow-minded and formal point of view they behave irrationally because they prefer the 0.99 probability of surviving in the high mountain to the sure surviving by staying home. In a similar position is an employee who leaves a large corporation to become an entrepreneur. Obviously, the rationality argument is weak here because it fails to capture essential aspects of human motivation.

A formal treatment of the two types of utility was suggested in early 21st century (Frey et al. 2004; Menestrel 2001). Le Menestrel (2001) introduced axioms accounting for both the utility of the future benefit as well as that of the actual process leading to it. From his standpoint, rational is the person who is optimizing not only the result, but also the experience.

A variation of experienced utility was studied by Benz (2007) who introduced the term *procedural utility* to denote the quality of the encounters one has with the institutions in one's capacity as citizen, customer, employee, tax-payer, plaintiff etc. Court decisions were protested less often, he noticed, when the court procedures were perceived as fair, impartial, and respecting the arguments of the two sides.

2.8 Rationality Principles and Their Violation

The attempt at describing human choice by the mathematical expectation principle was found problematic in the case of the St. Petersburg game, which gave rise to a paradox. Daniel Bernoulli introduced expected utility and thus resolved the issue, at the same time making an important step in quantifying a feature of the subjective human nature. The theory's further development in the 20th century caused great interest that helped refine it, but at the same time drew the borders beyond which it was shown to be inaccurate. Allais' paradox served as the cornerstone there. The majority of mainstream economists disregarded this finding for about a quarter of a century, until in 1979 prospect theory provided a convincing explanation to it. Meanwhile, expected utility theory continued to develop in several directions, among them the introduction of subjective probability by Savage.

However, new paradoxes were discovered. In 1961, Ellsberg (1961) came across another violation of the tenets of the theory, this time of its subjective variant introduced by Savage. (Some historians have claimed that Ellsberg in fact revived an idea initially proposed by John Maynard Keynes about four decades earlier.) In essence, subjective expected utility theory considered the subjective probabilities as additive, and now that was found to be contradicting the actual behaviour of many people. To be more precise, for incompatible events A and B, i.e. $A \cap B = \varnothing$, the following equality about classical probabilities is true:

$$p(A \cup B) = p(A) + p(B).$$

It turned out from Ellsberg's work, however, that for subjective probabilities such an equality does not hold, and

$$\pi(A \cup B) \neq \pi(A) + \pi(B). \tag{2.17}$$

In particular, Ellsberg devised an experiment with coloured balls drawn from urns, and asked his subjects about their choices. For concrete events A, B, and C people en masse reacted in ways that can be mathematically described with these inequalities:

$$\pi(A) > \pi(B) \tag{2.18}$$

and

$$\pi(A \cup C) < \pi(B \cup C). \tag{2.19}$$

If Eq. (2.17) were equality, then Eq. (2.19) would imply $\pi(A) < \pi(B)$. That, however, cannot be simultaneously true with Eq. (2.18). It follows, therefore, that the probabilities people employ in their decisions are not additive. Subsequently, other theorists have introduced modifications overcoming the problem and resolving the paradox.

At present, it is hard to say how many theories explain which paradoxes regarding economic choice. Perhaps the best source for orientation is a set of publications by Birnbaum (1999, 2004, 2008). He outlined a number of most established and simple postulates, and studied how the influential decision theories relate to them. Let us have a look at his classification.

1. *Transitivity.* This principle coincides with Axiom <u>A1b</u> from the von Neumann–Morgenstern axiomatic system.
2. *Coalescing.* This means that prospects of the type $(A, p_1; A, p_2; B, p_3)$ are unconsciously simplified to become $(A, p_1 + p_2; B, p_3)$. For example, a businessman is facing the alternative (\$50,000:50 %; \$50,000:10 %; \$0:40 %). This may be a competition for \$50,000 offered by a small municipality to do some construction works. The contract can be won by a number of competitors, each with different probability for success. In that case, the businessperson can submit two separate offers via two different firms, both owned by him. Intuitively, he might reformulate for himself the situation and arrive at the following representation: (\$50,000:60 %; \$0:40 %). Now he will assess his chances using that particular form.
3. *Branch Independence.* If two alternatives contain a common element, it is disregarded when they are compared. For example, if a choice must be made between $(A, p_1; B, p_2; C, 1 - p_1 - p_2)$ and $(A, p_1; D, q_1; E, 1 - p_1 - q_1)$, they are simplified to become $(B, p_2; C, 1 - p_1 - p_2)$ and $(D, q_1; E, 1 - p_1 - q_1)$.

According to Birnbaum, these three principles underlie all others, including the stochastic dominance, on which I will comment shortly. Discussing Allais' paradox, Birnbaum talked about which theories explain it by retaining or violating the above postulates:

> One class of theories (including subjectively weighted utility theory and original prospect theory) retains branch independence but violates coalescing, and thereby violates stochastic dominance. Another class of theories (rank-dependent and rank- and sign-dependent utility theories including cumulative prospect theory) retains coalescing and stochastic dominance but violates branch independence. (Birnbaum 1999)

This quote from 1999 illustrates well the state of fragmentation in which decision analysis found itself at the time. Decades later, new theories, postulates, and conjectures have explained old paradoxes and have stumbled into new problems. However common it may be for mathematicians to introduce variations in their axiomatic systems to examine the implications, that approach may not be entirely useful for studying human behaviour. This can be understood for example, by analogy with the engineering sciences. What would happen if in machine building, architecture, or civil engineering the axioms of Euclid be reshuffled in the way Birnbaum describes above, just to suit the needs of any particular task?

2.8.1 Rational Behaviour in Context

In 2006, Jörg Rieskamp, Jerome Busemeyer, and Barbara Mellers reviewed the rationality principles and put them in the natural context of adaptive human behaviour. They demonstrated that the old interpretation of rational choice as revealing internal individual preferences is unsatisfactory. In contrast, viewing choice at least as much as the product of environmental demands and circumstances, makes seemingly contradictory or suboptimal decisions look perfectly rational. Summarizing decades of research, these authors came up with a hierarchy of postulates wherein each implies the subsequent ones, with some exceptions. The important question though, is to what extent real behaviour complies with them. These are the principles:

1. Consistency of choice (Invariance of preferences across situations)
2. Strong Stochastic Transitivity
3. Independence from Irrelevant Alternatives
4. Regularity
5. Weak Stochastic transitivity.

Let me review the principles in relation with the ways in which people violate them. *Consistency of* Choice demands that a person maintain their preferences, e.g. $A \succ B$ under all circumstances. A related postulate is transitivity: $A \succ B$ and $B \succ C$ imply $A \succ C$. Numerous laboratory experiments and field studies have discovered many violations of this demand for consistency. But is it irrational?

Simon (1955, 1978) was among the first to object to such a conclusion. In his view, when the situation demands a quick decision, it might be far more advantageous to react quickly, possibly at the expense of some inconsistency. Just as important is another objection: in a dynamically changing environment with new opportunities emerging every now and again, the decision maker might benefit more by experimenting with them, rather than sticking to her/his established preferences.

Strong Stochastic Transitivity is a postulate that in the mathematical sense weakens the requirement for consistency of choice. Since people cannot be consistent across all situations, perhaps they could be "roughly" consistent. That means keeping preferences stable if not always, then most of the time. Let $p(A|\{A, B\})$ be the probability of choosing A among A and B. Instead of judging people's preferences by observing their inconsistent choices, we might better be interested in their probabilistic decisions: do they prefer A to B in more than 50 % of cases, i.e.

$$p(A|\{A,B\}) \geq 0.5.$$

Formally, Strong Stochastic Transitivity for each three alternatives A, B, and C means that if

$$p(A|\{A,B\}) \geq 0.5,$$
$$p(B|\{B,C\}) \geq 0.5$$

hold simultaneously, then

$$p(A|\{A,C\}) \geq \max[p(A|\{A,B\}), p(B|\{B,C\})].$$

This inequality is easily understood when one considers the following example. If A is preferred to B in 60 % of the cases when the choice is between them, and B is preferred to C in 90 % of the cases, then A must be preferred to C in 90 % of the cases when the choice is between A and C.

Since the 1950s, numerous experiments have shown massive violations of Strong Stochastic Transitivity. Rieskamp et al. (2006) cite the established psychological explanation of this effect, which is that in direct comparisons people pay disproportionate attention to the aspects in which the alternatives are different, but underestimate the common features among them. This mechanism distorts the objective picture and leads to violations of the postulate.

The next principle, *Independence from Irrelevant Alternatives*, requires that the preference between A and B should not change when each of them is evaluated with respect to option C or D. Formally,

$$p(A|\{A,C\}) \geq p(B|\{B,C\})] \Rightarrow p(A|\{A,D\}) \geq p(B|\{B,D\})].$$

It has been proven (Tversky and Russo 1969) that the most popular version of this principle is equivalent to the previous one, Strong Stochastic Transitivity. In that context, or outside of it, the principle of Independence from Irrelevant Alternatives is also violated en masse.

Weak Stochastic Transitivity is a postulate demanding that if *A* is preferred to *B* in over 50 % of the cases, and *B* is preferred to *C* also in over 50 % of the cases, then in direct comparison, *A* will be chosen over *C* in more than 50 % of all cases. There is a lot of experimental evidence that people generally behave in line with this postulate, with violations only rarely occurring.

Tversky, however, studied how and when consistent and predictable preference intransitivities, i.e., violations of the principle, occur. He showed that when people face five alternatives such as those in Table 2.2, their choices usually are as follows: $A \succ B, B \succ C, C \succ D, D \succ E$. Instead of the expected overwhelming dominance of *A* over *E*, however, very often people choose the opposite: $E \succ A$.

This decision anomaly relates to the effect of the just noticeable difference discussed in Chap. 1 and in Sect. 2.2, regarding axioms A2c and A2d and small probabilities. People hardly perceive the differences in probability between any two adjacent alternatives, and their attention focuses on the monetary outcomes. However, that is no longer the case when *A* and *E* get compared, and now probability begins to matter so much that people react like this: $p(E|\{A, E\}) \geq 0.5$.

Regularity is the last principle from the classification of Rieskamp and colleagues. It states that adding an option to a set of options can never increase the probability of selecting one of the initial options. Let $n > m$ and let $\Omega_1 = \{A_1,...,A_m\}$ be the initial set, to which new elements are added, to obtain $\Omega_2 = \{A_1, ..., A_m, A_{m+1}, ..., A_n\}$. Let A_i be an alternative from Ω_1 with probability to be chosen $p(A_i|\Omega_1)$. It is mathematically impossible to increase the chance for selecting A_i when new (competing) elements are added to the initial set. Therefore,

$$p(A_i|\Omega_1) \geq p(A_i|\Omega_2). \tag{2.20}$$

People may behave differently, however. If a newly added alternative is similar but worse than a particular alternative A_i from the original set, then the new one may actually increase the attractiveness of the original. For example, choosing a laptop between *A* and *B* may look like the case in Table 2.3. With all other features being identical, the only real differences are in weight and screen size.

If the customer's attention is drawn to a third laptop *C* with features as in Table 2.4, then option *B* suddenly rises in attractiveness. That is exactly the case when the inequality from Eq. (2.20) remains mathematically correct, but becomes descriptively wrong.

Table 2.2 Probabilistic alternatives

	A	B	C	D	E
Probability	7/24	8/24	9/24	10/24	11/24
Gain	$5.00	$4.75	$4.50	$4.25	$4.00

Table 2.3 Features of two laptops

	A	B
Weight (kg)	3	1.5
Monitor size (diagonal) (in.)	15	13

Table 2.4 Features of three laptops

	A	B	C
Weight (kg)	3	1.5	1.8
Monitor size (diagonal) (in.)	15	13	13

Sen (1993, 1995, 1997) has added an important argument to the adequacy of violating the regularity principle. Under certain conditions—mostly of social nature—almost anybody will act "irrationally", as in the following example. In a thought experiment, someone would much appreciate taking apple X from the fruit basket, but would refrain to do so if the apple was the last remaining. She would gladly take it though, if there were two remaining: X and Y. Let us define the following events: $A = \{Take\ X\}$; $B = \{Take\ nothing\}$; $C = \{Take\ Y\}$. Then the two sets are:

$$\Omega_1 = \{Take\ X; Take\ nothing\}$$
$$\Omega_2 = \{Take\ X; Take\ Y; Take\ nothing\}.$$

Good manners will lead quite a lot of people to violate rationality (regularity) and be described by

$$p(A|\Omega_1) < p(A|\Omega_2),$$

although $\Omega_1 \subset \Omega_2$.

Beyond the social norms, inducing people to behave in such a way, there are a number of instances where no such clear-cut arguments exist. In other words, regularity may be violated under a wider range of circumstances. A psychological theory that explains this behaviour is Busemeyer and Townsend's (1993) Decision Field Theory. It posits that the mental mechanism of choice involves a dynamic process of retrieving, comparing, and integrating prospect utilities U_t, where t indicates time. It starts with a deliberation phase whereby each alternative A_i is associated with utility that is a random variable at any moment in time. All such quantities $U_t(A_i)$ are integrated and form dynamic states, in which the mind focuses on the various aspects of the options at hand and the decision to be taken. Alternatives compete with one another and the more similar they are, the more intense is the competition among them. In that case, the deliberation takes longer and violations of the regularity principle increase in number. In contrast, it has been shown in numerous studies (see for example Todorov et al. 2005) that rapid unreflective inferences can be very powerful in the decision-making process.

2.8.2 The Dire Straits of Decision Analysis

It seems that the rigorous mathematical approach to decision analysis has encountered unexpected difficulties. The idea that guided it all along—the maximization of subjective utility—has brought some rewards, but has remained just as elusive. In incremental steps, it was reformulated in line with the scientific concepts and methods of each time period. However, the available mathematical techniques, and in more recent time the computer modelling tools have always been helpful, but were never the single decisive factor for understanding people's decisions. Ever since Bernoulli, this science has advanced only when it has incorporated new knowledge about how the human mind works, with formal analysis playing second fiddle.

Influential economists have recognized the limits of the approach seeking only internal consistency of choice. As Amartya Sen had put it, this very idea was "confused" because,

> [...] there is no way of determining whether a choice function is consistent or not without referring to something external to choice behaviour (such as objectives, values, or norms). We have to re-examine the robustness of the standard results in this light. (Sen 1993)

In other words, preferences hardly abide by logical consistency, but surely depend on the personality and motivation of the decision maker, who is also influenced by circumstances and context, and may or may not remain faithful to the urge for maximizing utility. This position frustrates some researchers because they find most valuable the knowledge that quantifies and ultimately predicts behaviour.

Whether science will ever reach such a level of pragmatic and efficient understanding is unknown. Yet, it is certain that there is a long way to go until we even begin to have satisfactory methods to meet such goals. Every now and again, we discover a new area for research, in this case the subject of human decisions, only to realize how ignorant we are.

References

Ahuvia, A. (2008). If money doesn't make us happy, why do we act as if it does? *Journal of Economic Psychology, 29*(4), 491–507.

Alchian, A. (1953). The meaning of utility measurement. *American Economic Review, 43*(1), 26–50.

Allais, M. (1953). Le comportement de l'homme rationnel devant le risque: Critique des postulats et axiomes de l'ecole americaine. *Econometrica, 21,* 503–546.

Allais, M., & Hagen, O. (Eds.). (1979). *Expected utility hypothesis and the Allais paradox.* Dordrecht: Reidel.

Arrow, K. (1965). *Aspects of the theory of risk bearing.* Helsinki: Yrjo Jahnsson Saatio.

Bell, D. E., Keeney, R. L., & Raiffa, H. (Eds.). (1977). *Conflicting objectives in decisions.* New York: Wiley.

Birnbaum, M. H. (1999). Paradoxes of Allais, stochastic dominance, and decision weights. In J. Shanteau, B. A. Mellers, & D. A. Schum (Eds.), *Decision science and technology: Reflections on the contributions of Ward Edwards* (pp. 27–52). Norwell, MA: Kluwer Academic Publishers.

Birnbaum, M. H. (2004). Decision and choice: Paradoxes of choice. In N. J. Smelser & P. B. Baltes (Eds.), *International encyclopaedia of the social & behavioural sciences* (pp. 3286–3291). Oxford: Elsevier.

Birnbaum, M. H. (2008). New paradoxes of risky decision making. *Psychological Review, 115,* 463–501.

Busemeyer, J. R., & Townsend, J. T. (1993). Decision field theory: A dynamic-cognitive approach to decision making in an uncertain environment. *Psychological Review, 100*(3), 432–459.

Dieffenbach, B. C. (1975). A quantitative theory of risk premiums on securities with an application to the term structure of interest rates. *Econometrica, 43*(3), 431–454.

Eeckhoudt, L., & Gollier, C. (1995). *Risk evaluation, management and sharing.* London: Harvester Wheatsheaf.

Ellsberg, D. (1961). Risk, ambiguity, and the savage axioms. *The Quarterly Journal of Economics, 75*(4), 643–669.

Fishburn, P. (1977) Book review on: Keeney, R. L., & Raiffa, H. (1976) *Decisions with multiple objectives: Preferences and value tradeoffs.* New York: Wiley. *Journal of the American Statistical Association, 72* (359), 683–684.

Fishburn, P. (1989). Foundations of decision analysis: Along the way. *Management Science, 35* (4), 387–405.

Frey, B., Benz, M., & Stutzer, A. (2004). Introducing procedural utility: Not only what, but also how matters. *Journal of Institutional and Theoretical Economics, 160,* 377–401.

Friedman, M., & Savage, L. J. (1948). The utility analysis of choices involving risk. *Journal of Political Economy, 56*(4), 279–304.

Kahneman, D., & Tversky, A. (1979). Prospect theory: An analysis of decision under risk. *Econometrica, 47*(2), 263–291.

Kahneman, D., & Tversky, A. (1984). Choices, values and frames. *American Psychologist, 39,* 341–350.

Keeney, R. L., & Raiffa, H. (1976). *Decisions with multiple objectives: Preferences and value trafeoffs.* New York: Wiley.

Lewin, S. B. (1996). Economics and psychology: Lessons for our own day from the early twentieth century. *Journal of Economic Literature, 34*(3), 1293–1323.

Machina, M. J. (1989). Dynamic consistency and non-expected utility models of choice under uncertainty. *Journal of Economic Literature, 37,* 1622–1668.

March, J. (1978). Bounded rationality, ambiguity, and the engineering of choice. *Bell Journal of Economics, 9,* 587–608.

Mas-Colell, A., Whinston, M. D., & Green, J. R. (1995). *Microeconomic theory.* Oxford: Oxford University Press.

Menestrel, Ml. (2001). A process approach to the utility for gambling. *Theory and Decision, 50,* 249–261.

Mosteller, F., & Nogee, P. (1951). An experimental measurement of utility. *The Journal of Political Economy, 59*(5), 371–404.

Pavlov, Y. P. (1989). A recurrent algorithm for the construction of the value function. *Comptes rendus de l'Academie bulgare des Sciences, 42*(7), 41–43.

Pavlov, Y. P. (2005). Subjective preferences, values and decisions. Stochastic approximation approach. *Comptes rendus de l Academie bulgare des Sciences, 58*(4), 367–372.

Pratt, J. (1964) Risk aversion in the small and in the large. Econometrica, *32*(1,2), 122–136.

Pratt, J. (1995). *Foreword* to Eeckhoudt, L., & Gollier, C. (1995). *Risk evaluation, management and sharing.* London: Harvester Wheatsheaf.

Preston, M. G., & Baratta, P. (1948). An experimental study of the auction-value of an uncertain outcome. *American Journal of Psychology, 61,* 183–193.

Rieskamp, J., Busemeyer, J. R., & Mellers, B. A. (2006). Extending the bounds of rationality: A review of research on preferential choice. *Journal of Economic Literature, 44,* 631–636.

Ross, S. A. (1981). Some stronger measures of risk aversion in the small and the large with applications. *Econometrica, 49*(3), 621–638.

Samuelson, P. (1938). A note on the pure theory of consumers behaviour. *Economica, 5*(17), 61–71.

Savage, L. J. (1954). *The foundations of statistics.* New York: Wiley.

Sen, A. (1993). Internal consistency of choice. *Econometrica, 61*(3), 495–521.

Sen, A. (1995). The formulation of rational choice. *The American Economic Review, 84*(2), 385–390.

Sen, A. (1997). Maximization and the act of choice. *Econometrica, 65*(4), 745–779.

Shafer, G. R. (1984) Comment to Allais, M., & Hagen, O. (Eds.). (1979). *Expected utility hypothesis and the Allais paradox.* Dordrecht, Reidel. *Journal of the American Statistical Association, 79*(385), 224–225.

Simon, H. A. (1955). A behavioural model of rational choice. *Quarterly Journal of Economics, 69,* 99–118.

Simon, H. A. (1978). Rational decision-making in business organizations. Nobel Memorial Lecture, 8 December 1978. *Economic Sciences,* 343–371.

Todorov, A., Mandisodza, A. N., Goren, A., & Hall, C. (2005). Inferences of competence from faces predict election outcomes. *Science, 308,* 1623–1626.

Tversky, A., & Kahneman, D. (1981). The framing of decisions and the psychology of choice. *Science, 211,* 453–458.

Tversky, A., & Kahneman, D. (1992). Advances in prospect theory: Cumulative representation of uncertainty. *Journal of Risk and Uncertainty, 5,* 297–323.

Tversky, A., & Russo, J. E. (1969). Substitutability and similarity in binary choices. *Journal of Mathematical Psychology, 6*(1), 1–12.

von Neumann, J., & Morgenstern, O. (1944, 1947, 1953) *Theory of games and economic behaviour.* Princeton: Princeton University Press.

Vorobyov, N. (1970) *Development of game theory.* Russian edition: von Neumann, J., and Morgenstern, O. (1944, 1947, 1953) *Theory of games and economic behaviour.* Princeton: Princeton University Press.

Weber, E. U., Blais, A. R., & Betz, N. (2002). A domain-specific risk attitude scale: Measuring risk perceptions and risk behaviours. *Journal of Behavioural Decision Making, 15,* 263–290.

Wilson, P. N., & Eidman, V. R. (1983). An empirical test of the interval approach for estimating risk preferences. *Western Journal of Agricultural Economics, 8*(2), 170–182.

Part II
Psychological Insights

Chapter 3
Prospect Theory—A Milestone

3.1 Two Cognitive Systems

Many theorists of economic rationality formulated insightful postulates with a lot of common sense, yet human decision-making proved a hard nut to crack. Even the mildest of all principles, weak stochastic transitivity, was—as we saw—proven to be systematically violated. The agents seem to be in so much contradiction with their own preferences, that virtually any attempt at understanding them has ended in frustration. Amartya Sen's call for major theoretical revisions was an expression at once of realism and probably some despair. Ironically, along the same decades, psychologists had entertained an idea that turned out to be of much help in decoding behaviour, and it was an idea so commonplace, that it could not be attributed to any particular researcher. Simply put, it stated that people approach all problems with one of two different modes of thinking—if the task is very simple, it is tackled swiftly and effortlessly, like the question how much is 2 + 2; however, tasks that are more complex demand not only more effort, but activate an altogether different mental machinery.

Figure 3.1 illustrates this point. In essence, information from the environment—physical and social—comes in through the senses and, if judged to be of no serious complexity and importance, prompts a quick decision from System 1 (Intuition), which handles most of the 'traffic' consisting of routine tasks. Should a complication arise, System 2 (Logical Reasoning) engages and deploys its sophisticated tools, which however, require a lot more cognitive effort.

Different theories have labelled the two cognitive systems in different ways, but in general it is believed that there is one system for "intuitive", "experiential", "impulsive", or "fast" reasoning, also called "System 1", and another for "logical", "rational", "reflective", or "slow" reasoning, also called "System 2" (Schneider and Shiffrin 1977; Strack and Deutsch 2004; Epstein 1994, 2003; Kahneman 2011; Kahneman and Frederick 2002; Stanovich and West 2000; Alós–Ferrer and Strack 2014; Brocas and Carrillo 2014; Dayan 2009). The intuitive system is automatic,

© Springer-Verlag Berlin Heidelberg 2015
G. Mengov, *Decision Science: A Human-Oriented Perspective*,
Intelligent Systems Reference Library 89, DOI 10.1007/978-3-662-47122-7_3

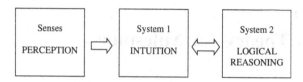

Fig. 3.1 Cognitive systems involved in decision-making

effortless, emotion-driven, governed by habit, but difficult to change, while the logical system is effortful, controlled and slow, but flexible and able to adopt complex decision rules. Easy tasks are dealt with predominantly by the former, while complications prompt the intervention of the latter. Choosing meals from a menu in a restaurant would demand mostly intuition—but being not entirely simple—would also need some input from the logical system, while working out the connections among the four alternatives in Allais' paradox would ask for a lot more of it. Looking it that way, Leonard Savage's U-turn at the famous 1952 Paris lunch is easy to understand: he was initially tricked into using his intuition, only to get alerted how insufficient it was.

In this view, the cognitive load due to every decision can be regarded as a point on a linear segment, with the domains of the intuitive and logical systems located at its ends. Of course, this dichotomy is somewhat schematic, as brain studies using modern scanning technologies have led to the understanding that the neural basis of emotion and cognition is highly integrated and hardly decomposable (Pessoa 2008). It has been suggested (Evans and Stanovich 2013) that it would be better not to use the "System 1" and "System 2" notions, but "Type 1" and "Type 2" *processes* instead, and hence, it would be more correct to refer to dual-process, rather than dual-system theories.

Somewhat apart from the family of dual-process theories is fuzzy-trace theory (Reyna and Brainerd 1991, 2008), which posits that intuition is "gist-based", i.e., resorting to vague memories about the gist of information relevant in a decision situation. In this view, intuition is more advanced than logical reasoning and is more characteristic of experts rather than novices. This is perhaps the only dual-system theory that regards intuition as a sophisticated form of mature reasoning. Indeed, a person trained to discover tricks like those in Allais' paradox would soon begin to do it automatically. The point is to receive training and develop expert intuition in relation to meaningful skills.

For their part, Tversky and Kahneman came to the idea of the two systems after an experiment with participants commanding statistical knowledge but failing to use it adequately in tasks demanding statistical intuition (Tversky and Kahneman 1971). Their finding suggested that making a quick judgement and using expert knowledge could involve different cognitive mechanisms. Processing of information from the senses (vision, hearing, touch, taste, and olfaction) with subsequent intuitive judgements and decisions is conducted with cognitive skills, developed earlier than logical reasoning in the evolutionary history of humanity.

While the senses continuously send information to the brain, it amplifies signals from the environment that reflect changes and differences. A process, which stays constant over some time receives diminishing attention as it has nothing new to say to the organism. This biological reaction is called *sensory adaptation*—through gradual attenuation of the response to the status quo, the nervous system remains alert to changes, as they could be potentially important. Such adaptation provides the *reference point* against which the new external influences are measured, and depends both on the previous and on the current stimulation levels. That is why identical stimuli trigger different reactions when the immediately preceding states have been different. A number of textbooks in introductory psychology give this example: A person holding her hands each in bowls of cold and warm water for some time, and then moving them both to a vessel with intermediary temperature, feels warm in one hand and cold in the other.

Similar things happen in social context. Drawing a biologically founded analogy between the system of perceptions and the intuitive system, Tversky and Kahneman postulated the existence of *adaptation level for welfare* as the decision maker evaluates the potential benefits from the available options. This is a radical departure from the classical model of utility theory. The tradition ever since Bernoulli had been to assess states of total wealth—if one chooses a risky alternative, one's final wealth W_f may turn out to be grater or smaller than one's present wealth W_0. Quantities W_f and W_0 are supposed to form the entire basis for the decision, but obviously, people do not behave like that. It is psychologically more realistic that one usually does not think of one's total assets, but rather about the change that would occur. In other words, the *gains* and *losses* are the important thing. Decision analysis has overlooked this effect for centuries and Kahneman (2003, 2011) dubbed it "Bernoulli's error".

3.2 Prospect Theory and Intuitive Thinking

3.2.1 Three Cognitive Features

Prospect theory is built around three tenets (Kahneman and Tversky 1979; Kahneman 2011) describing three essential cognitive features of decision-making. The first is the already discussed adaptation level for welfare, also known as *reference point*. It can be understood as the mix of current assets, income, expenses, habits with regard to luxurious or prestigious spending, or quality of life in general, to which the individual is accustomed. Any achievement below or above it is considered a loss or a gain. In addition, an entrepreneur or manager would focus on revenues, profits, and costs, which presumably comprise an important ingredient of their psychological wellbeing. Most often, the reference point coincides with the status quo, but there are exceptions. An entrepreneur or investor suffering a loss may notice that their industry in general is doing even worse, and therefore they perform better than expected. The industry is the reference point, the difference between it and their smaller loss can be viewed as a kind of gain. Another example would be a recent change in welfare (in any direction) that might have installed a new status quo, to which the agent has not yet adapted and would adhere to the preceding reference point for some time. Behavioural economist Matthew Rabin believes that the reference point is virtually always the level of welfare that the agent aims to achieve.

The second tenet, or principle, is *diminishing sensitivity to changes of wealth* (Kahneman and Tversky 1979; Kahneman 2011). This is an extension of the Weber-Fechner Law, applied to the more abstract domain of social life. Indeed, an increment in a gain (or loss) from $100 to $200 is more important to the individual than an increment from $1100 to $1200, exactly as it would be with differences among physical stimuli such as light, noise, heat, weight etc.

Finally, ingenious experiments have shown that people are much more sensitive to losses than to gains. That is why offers for gambles of the kind $C \equiv (-\$1000, 50\%; \$1500, 50\%)$, although apparently advantageous, are generally turned down. This principle has been called *loss aversion*. Such behaviour has evolutionary value—to survive and reproduce, organisms must treat potential threats as more important than potentially favourable opportunities (Kahneman 2011).

In his 2011 book (Kahneman 2011), Kahneman pointed out that these principles characterize System 1. The latter works together closely with System 2 to form the complex behaviour, observed in reality. For example, the above gamble C prompts System 1 to produce a negative emotion, which is the reason for System 2 to turn down the offer.

Figure 3.2 represents jointly the three principles. Utility is defined for changes in welfare, shown as shifts from the (neutral) reference point. The steepest regions of the graph are around the origin, where the ratio $u(x)/x$ is largest. This means that small gains have the relatively largest effect in terms of utility. In other words, happiness—when understood as the effect of positive emotion due to some gain—is

Fig. 3.2 Plot of the utility of a single outcome, considered as gain or loss

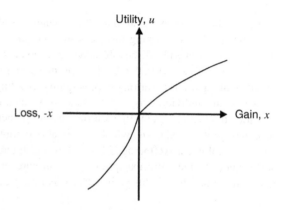

mostly about the small joys of daily life. Unfortunately, as the graph is much steeper in the negative domain, the smallest losses are the relatively most effective irritants that provoke negative affect.

Box 3.1   The Markowitz utility curve
In a 1952 article entitled "The Utility of Wealth" (Markowitz 1952), Harry Markowitz introduced a utility curve, similar in shape to that of prospect theory. It consisted of a mix of concave and convex regions and shared many features with the one proposed later by Kahneman and Tversky, including greater steepness for losses than for gains, and even associating utility with gains and losses rather than with total wealth. Markowitz, however, published his work before Allais' paradox became known, and his theory could not account for the violations of the expected utility principle.

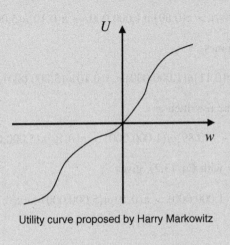

Utility curve proposed by Harry Markowitz

3.2.2 A Utility Function with Weighted Probabilities

Prospect theory modified the utility function of classical utility theory by introducing a kind of subjective probabilities, or "decision weights" as they were called at the time. The aim was to account better for the distortions of probability carried out by the human mind when assessing risky options. Essentially, the theory built upon the ideas of Ramsey and Savage, and on the empirical findings of Preston and Baratta, Mosteller and Nogee, Edwards, and others. For a two-outcome prospect, the new utility function looks like Eq. (2.3), shown in Sect. 2.4. Having terms like $\pi(p)u(x)$ instead of $p.u(x)$ weakens the principle of mathematical expectation and introduces more flexibility. Now Allais' paradox can be easily dealt with.

In Chap. 2, we observed a situation where people were unwilling to choose a very attractive prospect $B1$ because it came with only 99 % probability to materialize positively. However, a similar problem with equally reduced chances for success caused an overwhelming preference reversal. Formally, the preferences were initially $A1 \succ B1$ and later became $A2 \prec B2$. Modelling the situation with Eqs. (2.1) and (2.2) was inadequate. Now, due to the new utility function the same preference relations can receive a different analytical interpretation that resolves the contradiction. The guiding principle is that decision weights $\pi(p)$ deviate from the mathematical probabilities p. However, it is natural to assume that both kinds coincide for the impossible and the certain events, i.e., $\pi(0) = 0$ and $\pi(1) = 1$.

Following the von Neumann–Morgenstern tradition, an Appendix to prospect theory (Kahneman and Tversky 1979) outlined the axiomatic frame needed for the preference relations to imply algebraic relations. Thus, $A1 \succ B1$ can be represented as

$$\pi(1)u(1{,}000{,}000) > \pi(0.89)u(1{,}000{,}000) + \pi(0.10)u(5{,}000{,}000), \qquad (3.1)$$

while $A2 \prec B2$ becomes

$$\pi(0.11)u(1{,}000{,}000) < \pi(0.10)u(5{,}000{,}000). \qquad (3.2)$$

Equation (3.1) can be rewritten as

$$[\pi(1) - \pi(0.89)]u(1{,}000{,}000) > \pi(0.10)u(5{,}000{,}000),$$

which, taken jointly with Eq. (3.2), gives

$$[\pi(1) - \pi(0.89)]u(1{,}000{,}000) > \pi(0.10)u(5{,}000{,}000) > \pi(0.11)u(1{,}000{,}000).$$
$$(3.3)$$

Ignoring the expression in the middle of the double inequality in Eq. (3.3) and cancelling $u(1{,}000{,}000)$, we get:

$$\pi(1) > \pi(0.11) + \pi(0.89). \qquad (3.4)$$

Equation (3.4) can be generalized to become

$$1 > \pi(p) + \pi(1 - p). \qquad (3.5)$$

Equation (3.5) implies, in Kahneman and Tversky's words (Kahneman and Tversky 1979), that "the sum of weights associated with complementary [probabilistic] events is typically less than the weight associated with the certain event". They labelled that property *subcertainty*. In simpler words, it means that people tend to *underestimate* the probabilities when these are substantial, as in the above example. Subcertainty explains the "paradox" in Allais' tasks: when the numbers—money and probabilities—can fit in a double inequality like that of Eq. (3.3), preferences will always be reversed. Thus, large probabilities are generally underestimated, however, this can be up to a point because on the other hand, $\pi(1) = 1$. Therefore, there must be a kind of "jump" near the 100 %, which expresses the mind's hesitation to tell the difference between extremely high probability and actual certainty.

Unlike large probabilities, small probabilities are often *overestimated*. Consider this example. People generally view as equally attractive the following two alternatives: $C1 \equiv (\$350, 1.)$ and $C2 \equiv (\$1000, 0.5)$, which fact may be expressed as $(\$350, 1.) \cong (\$1000, 0.5)$. At the same time, when the probabilities for gains are reduced a hundredfold, the following preference is overwhelmingly observed: $(\$350, 0.01; \$0, 0.99) \prec (\$1000, 0.005; \$0, 0.995)$. These two statements can be combined as follows. Let $p = 1$, $q = 1/2$, $r = 1/100$, and also $x_1 = 350$, while $x_2 = 1000$. Then the equal preference implies

$$\pi(p)u(x_1) = \pi(pq)u(x_2).$$

Favouring the alternative with the larger sum when the probabilities have been reduced means that

$$\pi(pr)u(x_1) < \pi(pqr)u(x_2).$$

Bearing in mind the assumption that the utilities of gains are always positive, and so are the decision weights, one can write:

$$\frac{u(x_1)}{u(x_2)} = \frac{\pi(pq)}{\pi(p)}. \tag{3.6}$$

and

$$\frac{u(x_1)}{u(x_2)} < \frac{\pi(pqr)}{\pi(pr)}. \tag{3.7}$$

Taken together, Eqs. (3.6)–(3.7) imply the left-most inequality in the series:

$$\frac{\pi(pq)}{\pi(p)} < \frac{\pi(pqr)}{\pi(pr)} < \ldots < 1. \tag{3.8}$$

Of course quantities p, q, r, x_1, and x_2 do not need to be bound by the numbers in the last example.

Figure 3.3 offers a graphical interpretation to Eq. (3.8). When the probabilities in two competing alternatives as in the example become smaller, the human mind finds it difficult to set them apart. A difference between 100 and 50 % is felt easily, while a difference between 1 and 0.5 % is hardly perceived at all. Further reductions make the two probabilities indistinguishable, and the fractions in Eq. (3.8) get

Fig. 3.3 Diminishing
sensitivity for small
probabilities

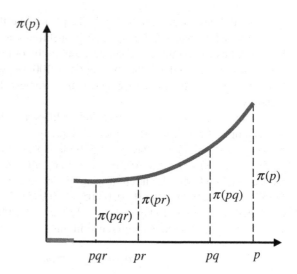

closer to one. This process continues, however, up to a point when the smallness becomes impossible to imagine, and the mind declares these chances equal to zero altogether. That is how intuitive thinking overestimates very small probabilities, but all of a sudden can neglect them and decide that the event is practically unfeasible.

In his book, Kahneman (2011) gave further details about when to expect overestimation of the chance for a rare event. It turns out that whenever the intuitive System 1 generates a vivid or emotional representation of the event, the latter is guaranteed to attract attention. Even discussing the event separately is enough to put it in focus. In such cases, overweighting is bound to happen. Otherwise, the rare event will be neglected and its chances will be considered equal to zero.

3.2.3 Cumulative Prospect Theory

Prospect theory envisaged an initial phase of decision-making during which all monetary outcomes, probabilities, and other relevant facts describing an alternative are reformulated, or "edited". In this way, they become suitable for utility to be quantitatively evaluated. During this phase numbers are rounded, identical outcomes with different probabilities get collated, common elements in the description of prospects are discarded to allow focusing on the differences, etc. A list of these and other operations was included in the original theory; it was, however, never meant to describe in detail how the human mind operates, but only to outline the general idea.

One such operation was the detection of dominance: when alternative A dominates another alternative B (i.e., all the outcomes of A are at least as good as those of B, with at least one outcome of A strictly better), B is quickly rejected from further evaluation. If the particular outcome is superior because of its higher probability of occurrence, this is called *stochastic dominance*.

Prospect theory assumed that during the editing phase the existence of dominance among alternatives is detected intuitively. This position, however, has provoked criticism on the account that the discontinuous weighting function, as was shown in Fig. 3.3, entails violations to the stochastic dominance principle. On the other hand, it has been mathematically proven that if a theoretical model abides by the stochastic dominance principle, its decision weighting function must be consistent with the problematic von Neumann–Morgenstern theory.

One way around the obstacle was to assume that decision weights are not associated with the isolated probability, but with the entire probability distribution (Quiggin 1982; Schmeidler 1989). Now some probabilities in a complex alternative might be overweighted at the expense of others. A reason for this to happen might be that very favourable outcomes naturally attract more attention, just like do the most undesirable ones. Take the example of a person evaluating the prospect ($300, 1/3; $200, 1/3; $100, 1/3). An inveterate pessimist, they may consign half of the importance to the least favourable outcome, $100. While the probability of each outcome is 1/3, the decision weights may be for instance $\pi_1 = \pi_2 = 0.25$, and $\pi_3 = 0.50$ (Diecidue and Wakker 2001). The assumption that the decision maker is ordering in their mind (assigning ranks to) the possible outcomes when assessing the entire prospect, has given rise to the name *rank-dependent* theories. Because decision weights now depend on the entire probability distributions, all such theoretical models are also called *cumulative*. A realistic treatment should be able to account for not only the distortion of probabilities, but also for the fact that there are better and worse outcomes relative to one another. Theorists such as Quiggin, Schmeidler, Yaari, Weymark and others proposed new analytical representations that generally became popular under the name rank-dependent utility theory. They all overcame Allais' paradox and other similar problems, but encountered another kind of difficulties.

A major issue comprises the separability of decision weights and utilities of outcomes. Mathematically, it makes sense to use a term like $\pi(p)u(x)$ only under the assumption that $\pi(p)$ and $u(x)$ are very much independent of each other. Defining the decision weights as a function of not only probabilities, but also of: (i) outcomes x; (ii) the outcome relative values with respect to one another; (iii) their being gain or loss; (iv) their number in a complex prospect; and finally (v) their verbal formulation would compromise the separability principle. Hence, all these factors would dictate abandoning the simple representation $\pi(p)u(x)$ in favour of more complex ones.

In a new variant of prospect theory, Tversky and Kahneman (1992) managed to find a middle ground between the conflicting demands of keeping stochastic dominance intact and maintaining the separability of weights from utilities. They did so by setting apart all positive and negative outcomes of a complex prospect, and then bundling the two sets together to form two sums of products as in

Eq. (3.9). The new cumulative prospect theory introduced the following utility function:

$$U(A) = \sum_{i=1}^{m} \pi_i^+ u(x_i) + \sum_{j=m+1}^{n} \pi_j^- u(x_j). \tag{3.9}$$

In Eq. (3.9), m is the number of outcomes related to gains, and $n - m$ is the number of outcomes bringing losses ($n, m \in N$, $n \geq m \geq 0$). Quantities π_i^+ and π_j^- are related to all positive and negative outcomes, respectively. Now, a decision weight is defined by Eq. (3.10) as the marginal contribution of the particular outcome. For example, the ith positive outcome has this weight:

$$\pi_i^+ = w^+(p_i + \ldots + p_m) - w^+(p_{i+1} + \ldots + p_m). \tag{3.10}$$

In Eq. (3.10), p_i are probabilities and w^+ is the weighting function associated with all positive outcomes. The new theory is at once rank-dependent (and retaining stochastic dominance) and keeping the decision weights sufficiently independent from the outcomes' utilities.

Empirical research has shown the following algebraic function to be a reasonably good fit for people's experimentally observed behaviour:

$$w(p) = \frac{p^\delta}{\left(p^\delta + (1-p)^\delta\right)^{1/\delta}}, \tag{3.11}$$

In Eq. (3.11), the parameter δ is determined empirically. Figure 3.4 presents the decision weighting functions for gains and losses introduced by cumulative prospect theory. Their curvature embodies an important principle, shared also by the utility function. The latter, as we saw in Fig. 3.2, has the property of diminishing sensitivity as one gets further away from the reference point. The same is true about decision weights now. There are two natural reference points with regard to probabilities: the certain event and the impossible event. Equation (3.11) ensures that weights are more sensitive to the probabilities near these two boundaries, and loose some sensitivity in-between. Exactly that is what we see in Fig. 3.4.

Having made the best use of the idea for rank-dependence, Tversky and Kahneman remained skeptical about its practical value (Tversky and Kahneman 1992). They were all too aware how important is the framing of alternatives and how crucial is the way emotions are provoked. After all, they had devised examples such as the Asian Disease Problem to demonstrate the instability of human preferences when primed by positive and negative formulations of a situation. In 2002, many decades after the two scientists' original study, a graduate student of mine, Elitsa Hristova, and I conducted a similar experiment with somewhat updated text (Mengov and Hristova 2004). It stated:

Fig. 3.4 Decision weights and probability, modelled by Eq. (3.11). *Thick line* weighting function for the probability of gains. *Thin line* weighting function for the probability of losses

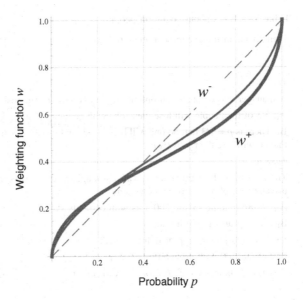

You are the boss of the Privatization Agency in an East European country. You examine two proposals from potential buyers of a state-owned company. At risk are 600 jobs. Your choice must be based solely on the buyers' programmes to save jobs. The consequences will be as follows:

If programme A is adopted, 200 jobs will be saved.
If programme B is adopted, there is 1/3 probability that all 600 jobs will be saved and 2/3 probability that all 600 people will be laid off.

We conducted a survey among a haphazard sample of Bulgarian economists to study their intuition with regard to such a socioeconomic situation. For a control group we used Bulgarian software engineers. We expected preference reversals on the scale of the original study, but hoped that the economists might do slightly better due to their education and professional experience. It was not to be. Table 3.1 shows the results we got.

Each participant read a questionnaire, constructed to present the two tasks about five to six minutes apart. As expected, this time span was enough to forget the first task. The emotional System 1 swept away economists and software engineers alike, prompting them to like saving jobs and hate lay-offs, at the same time ignoring the equivalence of the outcomes. As Table 3.1 shows, less than half of all people gave consistent responses. This was another example for what Kahneman (2011) dubbed "empty intuition".

Even without emotional descriptions but only using tasks formulated as "Which option do you prefer: $(x_1, p; 0, 1 - p)$ or $(x_2, 100\%)$?" one can elicit substantial instability of preferences. In another experiment, I gave to subjects a sequence of such simple alternatives to compare. A typical performance is shown in Fig. 3.5.

Table 3.1 Tasks about laying off people

Participants in the experiment about 600 jobs	Economists (n = 52)	Software engineers (n = 48)
Task I		
A manager chooses between options A_I and B_I how to cut costs by laying off people		
A_I: 2/3 of the people will lose their jobs with certainty (%)	38	42
B_I: There is $p = 1/3$ that no one will be laid off, and $p = 2/3$ that all will lose their jobs (%)	62	58
Task II		
The boss of an East European privatization agency chooses between options A_{II} and B_{II} for privatization of a state-owned company		
A_{II}: A programme saving 200 jobs with certainty (%)	87	60
B_{II}: A programme offering to:	13	40
Save all 600 jobs with probability $p = 1/3$		
Lay off all 600 with probability $p = 2/3$		
Consistent choices (A_I & A_{II}) or (B_I & B_{II}) (%)	48	48

Fig. 3.5 Empirically obtained decision weights from a participant in an experiment. The fitting line is computed with Eq. (3.11)

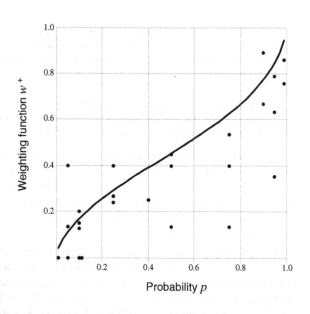

There, we see expected underestimation of moderate to large probabilities, while probabilities of 5 and 10 % are either reduced to zero, or taken to reach as high as 20 % or even 40 %. The curved line is computed with Eq. (3.11). It seems that choice instability is an inherent characteristic of our cognitive apparatus.

Cumulative prospect theory introduced not only an analytical representation for the decision weights, but also one for the utility of single outcomes. Careful experimenting led to the following formula for the outcome in a risky prospect:

$$u(x) = \begin{cases} x^{\alpha} & \text{if } x \geq 0 \\ -\lambda(-x)^{\beta} & \text{if } x < 0. \end{cases} \tag{3.12}$$

Constants α and β in (3.12) are real numbers within (0, 1). The difference between them was found to be statistically insignificant, suggesting that the mind's sensitivity for gains and loses diminishes at the same rate. Parameter λ is responsible for the difference in steepness of the utility function in the two domains as was shown in Fig. 3.2. It is a measure of the greater emotional intensity due to the pain from losing an amount x as compared with the joy from receiving x. That parameter was empirically assessed (Tversky and Kahneman 1992) to be equal to 2.25. In other experiments it varied in the range of 1.5 and 2.5 (Kahneman 2011). While this quantity is in no way on par with the constants in the natural sciences, it provides a good orientation about the mind's emotional response. In practical terms, it means that if something valuable is lost, the person losing it must receive at least a twofold compensation to recover their initial wellbeing.

Equations (3.9)–(3.12) are sufficient to summarize a huge amount of choice variability in situations under risk and uncertainty. The two functions in the product $\pi_i u(x_i)$ form a reasonably good model for what is happening in one's mind when one faces probabilistic alternatives. Tversky and Kahneman called it the *fourfold pattern* of human behaviour. Its first element is this. Because high probabilities are underestimated, one is willing to receive less money than the mathematical expectation for a risky prospect, especially when the potential gain is large and is perceived to be much greater than the ordinary amounts one is accustomed to. This happens just like in the St. Petersburg's game. The second element of the fourfold pattern is observed when the same high or at least substantial probability comes together with a potential loss. Probability is again underestimated, but this time it helps to induce bold and daring behaviour. Finally, the last two patterns concern the disproportionate influence of small probabilities on choices due to their huge decision weights. They encourage people to spend money to avoid risks by buying insurance policies; for the same reason they are the motive behind buying lottery tickets. That is how cumulative prospect theory settled the intriguing question by Friedman and Savage about why people are willing to sell and buy risks at the same time. In addition, detailed experimenting indicated that, "…there is no general trait of risk aversion or risk seeking" (Tversky and Kahneman 1992). This conclusion is very much in line with the lack of consistent attitude towards risk, discussed already in Chap. 2.

3.3 Methodological Issues

Prospect theory was a huge success. It attracted much attention and was cited a lot. It brought a Nobel Prize for economics to a psychologist—a remarkable achievement in its own right. In his book published almost a decade after the prize, Kahneman shared his thoughts about the reasons for this good fortune (Kahneman 2011):

> ...Then we constructed a theory that modified expected utility theory just enough to explain our collection of observations. Prospect theory is ... more complex than utility theory. In science complexity is considered a cost, which must be justified by a sufficiently rich set of new and (preferably) interesting predictions of facts that the existing theory cannot explain. Richer and more realistic assumptions do not suffice to make a theory successful. Scientists use theories as a bag of working tools, and they will not take on the burden of a heavier bag unless the new tools are very useful. Prospect theory was accepted by many scholars not because it is "true" but because the concepts that it added to utility theory, notably the reference point and loss aversion, were worth the trouble; they yielded new predictions that turned out to be true.

This position is remarkable on a number of levels and deserves to be discussed at each of them. To begin with, it is a lesson about how someone so successful can remain so modest. As science developed over the centuries, its history recorded a large variability of reactions to one's own success. Here we have an example worthy of genuine respect. Further, Kahneman was not ashamed to admit that his and Tversky's theory had some "blind spots": for example,—just like utility theory—it could not deal with disappointment and regret.

Introducing in a controlled way complexity in the theoretical model of expected utility was a crafty strategy. When empirical data guides the theorizing process and not the other way round, sometimes a researcher can remain more open to the signals emitted by the object of investigation. Such approach may be doubly justified if the object is as complicated as human behaviour is. The new theory did not lack in rigour either. It is often forgotten that prospect theory had an axiomatic foundation no less solid than that of utility theory. This foundation, however, was never given centre stage and was in fact relegated to an appendix in the original article (Kahneman and Tversky 1979). Today the theory is remembered probably more than anything else for the variety of interesting psychological effects, relevant to economic decision-making, that were discovered alongside its development.

On the other hand, building up prospect theory by following utility theory so closely could not bring advantages only. The above "blind spots" are perhaps the lesser issue. Far more important is that for all its brilliance and fame, Kahneman and Tversky's construction was a brainchild of its time and the limitations of its time's scientific methods and instruments, which confined it to become a purely phenomenological[1] theory. Due to its roots, therefore, it could not explain any of

---

[1]"Phenomenological" has different meanings in philosophy, psychology, and the natural sciences. Here I use the term as it is understood in the latter sciences, most often in physics: it qualifies theories and mathematical models that fit empirical data without relating to the deep mechanism of the phenomena.

the intricate cognitive mechanisms driving human decisions. Its authors were forced to maneuver and trade off one postulate for another in the pursuit for better modelling the effects they discovered. This is how they themselves viewed the situation:

> ...the cumulative version – unlike the original one – satisfies stochastic dominance. Thus, it is no longer necessary to assume that transparently dominated prospects are eliminated in the editing phase... On the other hand, the present [cumulative] version can no longer explain violations of stochastic dominance in nontransparent contexts. (Tversky and Kahneman 1992).

An unhappy consequence of this arrangement was the sacrifice of the subproportionality property. In the original theory, it characterized the mind's diminishing sensitivity for small probabilities as illustrated by Fig. 3.3. The cumulative version resorted to the opposite principle: weights are more sensitive to probabilities near the two boundaries of zero and one, and lose sensitivity in-between, as seen from Fig. 3.4 and Eq. (3.11). Of course, both principles are backed by common sense while the empirical evidence (for example, Fig. 3.5) is so inconclusive that can serve them equally well. Yet, scientifically this is hardly a consolation. Prospect theory, in both its denominations, remains a good description for tendencies and statistics but should not be summoned to account for every observed detail. In retrospect, introducing the more complex mathematical apparatus of rank-dependent models does not seem to have paid off because as much explanatory power was gained in one direction, was lost in another. It looks as if decision science—in the hands of two of its leaders—was experimenting with various analytical tools and was seeking the most promising path forward, only to remain in hesitation.

Years later, after new studies had uncovered new cognitive phenomena, Birnbaum (2008) compiled a list of 11 paradoxes that both versions of prospect theory could not explain. His alternative TAX (Transfer of Attention Exchange) theory (Birnbaum 2008, 2004; Birnbaum and Chavez 1997)—a more general rank-dependent construction—claimed to be able to account for them all. Because it is more complex, it is indeed able to explain a larger amount of empirical data. However, it is just as phenomenological. Should it be lucky to draw more attention, it seems inevitable that some ingenious experiment would find fault with it just at it has happened with many theories before.

Finally, let me come back to the virtue of simplicity in a theory as exemplified by the one discussed here. It seems fair to say that the precision with which Tversky and Kahneman have used Occam's razor deserves a place in textbooks on methodology of science. It is hard to see how one could have accomplished more. Yet the observer is left feeling uneasy about at least two controversial points. The first is this: Once utility theory was known to be so flawed, and was further found to be deficient in important new ways, why was it treated with the ultimate respect of being used as the mould for the new theory? One may only try to guess the answer: Was there a lack of sensible alternative strategies? Or was there a somewhat excessive regard for tradition? On the other hand, maybe it was tempting to take the economists' community gently by the hand and teach them a couple of things about the psychology of economic decision-making?

The second point of contention is related. Incremental steps are valued in all fields of research. However, anybody working in the natural sciences, engineering, or computer science would be puzzled to learn that a theory is accepted, *"because the concepts it added ...were worth the trouble"*. After Thomas Kuhn's paradigm shifts and Imre Lakatos' research programmes, scientists are open to a lot of subjectivism, but still hope for some less emotional style of assessment. Perhaps over time decision science could converge to the natural sciences in the pursuit for more solid first principles. The message in the cited words can be accepted only by acknowledging that in his modesty, Kahneman also winks at the readers with a hue of his light humour.

The book *Thinking, Fast and Slow* was praised as an achievement equaling Adam Smith's *The Wealth of Nations*. Indeed, it is not only useful and instructive but also makes a pleasant reading. Still, this "lifetime's worth of wisdom"—in the words of Steven Levitt—moves the reader not so much with its theoretical constructions but mostly due to the multitude of diverse psychological facts that blend in an innovative but realistic account of human behaviour. This is the mastery of Kahneman—to derive wisdom from experiments, based on the limited scientific methods of his time, and somehow manage to stand above the details and paint a captivating bigger picture.

References

Alós–Ferrer, C., & Strack, F. (2014). From dual processes to multiple selves: Implications for economic behaviour. *Journal of Economic Psychology, 41*, 1–11.

Birnbaum, M. H. (2004). Decision and choice: Paradoxes of choice. In N. J. Smelser & P. B. Baltes (Eds.), *International Encyclopaedia of the social & behavioural sciences* (pp. 3286–3291). Oxford: Elsevier.

Birnbaum, M. H. (2008). New paradoxes of risky decision making. *Psychological Review, 115*, 463–501.

Birnbaum, M. H., & Chavez, A. (1997). Tests of theories of decision making: Violations of branch independence and distribution independence. *Organizational Behaviour and Human Decision Processes, 71*, 161–194.

Brocas, I., & Carrillo, J. D. (2014). Dual-process theories of decision-making: A selective survey. *Journal of Economic Psychology, 41*, 45–54.

Dayan, P. (2009). Goal-directed control and its antipodes. *Neural Networks, 22*, 213–219.

Diecidue, E., & Wakker, P. P. (2001). On the intuition of rank-dependent utility. *The Journal of Risk and Uncertainty, 23*(3), 281–298.

Epstein, S. (1994). Integration of the cognitive and the psychodynamic unconscious. *American Psychologist, 49*(8), 709–724.

Epstein, S. (2003). Cognitive-experiential self-theory of personality. In T. Millon & M. J. Lerner (Eds.), *Comprehensive handbook of psychology* (Vol. 5, pp. 159–184). Personality and Social Psychology Hoboken, NJ: Wiley.

Evans, J. St. B. T., & Stanovich, K. E. (2013). Dual-process theories of higher cognition: Advancing the debate. *Perspectives on Psychological Science, 8*(3), 223–241.

Kahneman, D. (2011). *Thinking, fast and slow.* NY, New York: Farrar, Straus, and Giroux.

Kahneman, D. (2003). Maps of bounded rationality: Psychology for behavioural economics. *The American Economic Review, 93*(5), 1449–1475.

Kahneman, D., & Frederick, S. (2002). Representativeness revisited: Attribute substitution in intuitive judgement. In T. Gilovich, D. Griffin, & D. Kahneman (Eds.), *Heuristics and Biases: The psychology of intuitive judgement*. New York: Cambridge University Press.

Kahneman, D., & Tversky, A. (1979). Prospect theory: An analysis of decision under risk. *Econometrica, 47*(2), 263–291.

Markowitz, H. (1952). The utility of wealth. *The Journal of Political Economy, 60*(2), 151–158.

Mengov, G. (2014). Person-by-person prediction of intuitive economic choice. *Neural Networks, 60*, 232–245.

Pessoa, L. (2008). On the relationship between emotion and cognition. *Nature Reviews Neuroscience, 8*, 148–158.

Quiggin, J. (1982). A theory of anticipated utility. *Journal of Economic Behavior and Organization, 3*, 323–343.

Reyna, V. F., & Brainerd, C. J. (1991). Fuzzy-trace theory and framing effects in choice: Gist extraction, truncation, and conversion. *Journal of Behaviour and Decision Making, 4*(4), 249–262.

Reyna, V. F., & Brainerd, C. J. (2008). Numeracy, ratio bias, and denominator neglect in judgments of risk and probability. *Learning and Individual Differences, 18*, 89–107.

Schmeidler, David. (1989). Subjective probability and expected utility without additivity. *Econometrica, 57*, 571–587.

Schneider, W., & Shiffrin, R. M. (1977). Controlled and automatic human information processing: 1. Detection, search, and attention. *Psychological Review, 84*, 1–66.

Stanovich, K. E., & West, R. F. (2000). Individual differences in reasoning: Implications for the rationality debate? *Behavioural and Brain Sciences, 23*, 645–665.

Strack, F., & Deutsch, R. (2004). Reflective and impulsive determinants of social behaviour. *Personality and Social Psychology Review, 8*, 220–247.

Tversky, A., & Kahneman, D. (1971). Belief in the law of small numbers. *Psychological Bulletin, 76*, 105–110.

Tversky, A., & Kahneman, D. (1992). Advances in prospect theory: Cumulative representation of uncertainty. *Journal of Risk and Uncertainty, 5*, 297–323.

Chapter 4
Intuitive Judgements

4.1 Analytical Means to Manage Uncertainty

Imagine that your job is to analyze data for your organization, which could be of any kind. Whether it is business data related to sales, revenue, market share, media coverage, production lines, agriculture, human resources, headcount, training, healthcare or something very different, you must understand what it is telling you and anticipate what trend might be expected. Your position—of a manager or an analyst—is remote enough from the 'front line' and you cannot rely too much on your intuition, gut feelings, or hands-on experience. All you have is numbers in a file that you can plot as dots in a picture like Fig. 4.1. Your education and additional training are a huge source of ideas, but none is concrete enough to give a straight answer to your question. If you are an engineer, you may be assisted by knowledge in physics, chemistry or some other natural science whose laws may have been harnessed in semi-empirical theories, algorithms, or procedures—yet the uncertainty at hand hinders any straightforward application of such approaches with regard to the particular dots. What could be done?

Simple linear regression seems to be a natural first step. Equation (4.1) below would give a rough estimate of the tendency. If the data are normally distributed and coefficients a_0, a_1 turn out to be statistically significant, and hopefully the variance of error ε is not too high, the approximation (4.1) would be tolerable for each particular dot, old and new:

$$f(x) = a_0 + a_1 x + \varepsilon. \tag{4.1}$$

Under the same assumptions, the parabola by Eq. (4.2) would be a better fit:

$$f(x) = a_0 + a_1 x + a_2 x^2 + \varepsilon \tag{4.2}$$

© Springer-Verlag Berlin Heidelberg 2015
G. Mengov, *Decision Science: A Human-Oriented Perspective*,
Intelligent Systems Reference Library 89, DOI 10.1007/978-3-662-47122-7_4

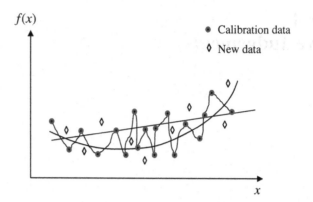

Fig. 4.1 Fitting empirical data with three different models. The *straight line* is the simplest model, but it is not accurate enough. An excellent fit for the calibration data, the *curved line* is a failure with all new data (the *rhombic dots*). A parabola offers a compromise for both the calibration and the test data: it is therefore better than the other two

With luck, coefficient a_2 and at least one of a_0, a_1 would be significant. Undoubtedly superior, however, would appear to be a model of this kind, assuming $m > 3$:

$$f(x) = \sum_{j=0}^{m-1} a_j x^j + \varepsilon. \tag{4.3}$$

Here, $m - 1$ is the degree of the polynomial and m is the number of parameters to be estimated. Now error ε would be approaching zero, as the curved line would go through each dot. However, almost every a_j would be statistically insignificant—a warning for imminent failure. Indeed, our goal was to use the current data to predict the new data—the rhombic dots—and the complicated model (4.3) is virtually useless for them. The naked eye is in doubt to favour the parabola over the straight line, or vice versa. Hypothetically, a simple computational procedure would confirm that the parabola is better.

We would like to know which model to prefer *before* the new data came. A definitive answer does not exist. The coefficient of determination R^2 by Eq. (4.4) cannot be of much help, but is a logical starting point for the discussion. It is defined as

$$R^2 = \frac{\sum_i (\hat{f}_i - \bar{f})^2}{\sum_i (f_i - \bar{f})^2}, \tag{4.4}$$

where f_i are the real observations ($i = 1, \ldots, n$), while $\hat{f}_i$ are their estimates by equations like (4.1–4.3) or any other, and $\bar{f}$ is the mean over all f_i. Equation (4.4) gives a measure of how close all values $\hat{f_i}$ are to the really observed f_i. Quantity R^2 varies within $0 \leq R^2 \leq 1$ and measures how correlated the real and the modelled data are. It would erroneously proclaim the most complex model (4.3) as the most suitable. In fact, the more complex a model is—in terms of number of parameters— the easier it is for it to capture more information from the data available for calibration. "Give me five parameters and I will describe a whole elephant with them", the great physicist Enrico Fermi is said to have exclaimed.

However, all measurements contain 'noise', i.e. influences by random factors that get absorbed by the parameters indiscriminately. Therefore, it would be wise to judge models by their ability to compress more information in as few parameters as possible. We could have some trust in a model only when $n \gg m$. A way to achieve a measure more accurate than R^2 would be to extend it by incorporating the parameters-to-observations ratio. The adjusted R^2 proposed by Weary has this form:

$$R^2_{adj(W)} = 1 - \frac{(n-1)}{(n-k-1)}(1 - R^2),\tag{4.5}$$

where k is the number of those among all m parameters, which are statistically significant. Now a large R^2 will often go along with a much smaller $R^2_{adj(W)}$, which would be a more realistic indicator for the quality of the model. Even better would be the Hertzberg variation:

$$R^2_{adj(H)} = 1 - \frac{(n-1)}{(n-k-1)}\frac{(n-2)}{(n-k-2)}\frac{(n+1)}{n}(1 - R^2).\tag{4.6}$$

Which is the best such measure? There is no principled way of telling. Yet another alternative is the Akaike Information Criterion (AIC):

$$AIC = \ln\frac{1 - R^2}{n} + 2m.\tag{4.7}$$

Now the goal is to have AIC as small as possible—as R^2 grows the *rhs* of Eq. (4.7) diminishes, but this may be happening due to very large number of parameters m. The second term $2m$ introduces a kind of penalty.

Another sensible idea is the "Jack-knife resampling": the model is estimated with all data records except one, which is used for prediction. Then it is put back in and another one is hold out, and so on until all n records are predicted. An average over all predictions would serve as an estimate for the model capability. Still, the method depends entirely on the available observations and gives no guarantee for success with unknown data.

While R^2 is a measure for the data variance captured by the model, $R^2_{adj(W)}$, $R^2_{adj(H)}$, and AIC aspire to assess how adequate a model would be not only with the data in hand, but with all other hypothetical samples from the same distribution. Because this is an impossible task, none of these measures is perfect. Many more have been introduced and they all bear the umbrella title Goodness-of-Fit Index (GFI). Various fields in the applied sciences have adopted different indices: Adjusted GFI (AGFI), Comparative Index of Fit (CFI), Parsimony Index Ratio (PIR), Chi-squared indices, including $AIC\chi^2$, and dozens of others.

All of these criteria seek to incorporate two obvious ideas. The first is that if a model is too simple it would fail to account for important informational aspects of the phenomenon. And as Einstein put it, "Everything should be made as simple as possible, but not simpler." The second idea is that a complex model will deal easily with the available (calibration) data but will fail with any new data, as we saw in Fig. 4.1. To William of Ockham, a 14th century Franciscan Friar belongs the most succinct formulation of this principle, known also as Ockham's Razor: "Do not add essences without necessity."

Starting in the 1960s, the Russian mathematicians Vladimir Vapnik and Alexey Chervonenkis developed statistical learning theory, which clarified the issue of the best model, at least at the theoretical level. They studied a class of models, or "learning machines" in their terminology, and established that the best generalization performance, i.e. the ability to account for unknown data is achieved when the "machine capacity" is made optimal for the available training data. Of course, the new data must not be statistically different from the available. In terms that are more common and with some simplification the idea can be explained as follows.

Take the example of Fig. 4.1 and Eqs. (4.1)–(4.3). Apparently, the latter are of the same kind and the last one is the most general, including the former two as special cases. As the models become more complex, they capture a larger amount of the information in the calibration sample. The error, understood as some integral measure (e.g. the mean squared error) of the discrepancy between observed and model-predicted data decreases monotonously. However, when the new data come, the most complicated equation may not produce the best fit. As we saw in the example, a model of middle complexity could be the most adequate. In the Vapnik–Chervonenkis theory, the model complexity is measured by a quantity called VC-dimension. For low-dimensional problems, it is approximately equal to the number of parameters to be estimated. Figure 4.2 suggests that for each problem there is an optimal model complexity. For new data, there is an upper bound ε_{test} for the prediction error that is nonlinear, having a minimum and then growing up again. A modeler should seek the optimal zone between the dotted lines and should not be tempted to "add essences without necessity".

Since the 1980s, the field of artificial neural networks quickly gained momentum in data modelling and prediction. Its most popular representative, the multilayered perceptron (MLP) was summoned to help in a multitude of problems. This field turned out to be an excellent territory for application of ideas from statistical learning theory. In particular, having in mind the concept of optimal machine capacity, researchers were better equipped to establish the best neural structure for each task they were dealing with. Further, it was discovered that a given network might perform differently with training and test samples if it received different amounts of training. A network might be structurally optimal, but if not trained long

Fig. 4.2 Theoretical error
curves in calibration and test
samples

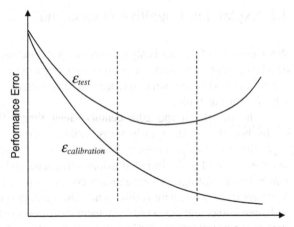

Calibration Effort (Training Time and/or Model Complexity)

enough might end in the undesirable zone to the left of the dotted lines in Fig. 4.2. An overtrained network, in contrast, might go too much to the right, doing "overfitting" like the curved line in Fig. 4.1. The optimal calibration strategy was soon discovered—a sufficiently large part of the training sample must be hold out as quasi-test data: It will help to detect in which part of the ε_{test} curve the network is currently to be found. Should this quasi-test error rise compared to previous runs, the model training has gone too far.

Now let us take some distance from the technical discussion and think again about quantitative analysis in everyday business. Most of the data people deal with is not as important as to demand the attention of bright scientists who would conduct an expensive research about it. Normally, we are facing routine tasks that need quick answers of moderate precision, for which methods like regression analysis, principle component analysis and other of the kind are good enough.

To borrow the idea about the intuitive vs. rational system dichotomy, it is as if any organization is a group agent with two such systems, this time operating not at the individual level, but on the scale of the entire collective. "Business as usual" should be more or less effortless and governed by habit. Only when the organization faces a serious problem, it deploys more costly procedures, invests more resources of every kind, and adopts complex decision strategies. Hence, by virtue of their routineness, the standard analytical approaches discussed in this section may be considered as an extension of human professional intuition. In this chain of thought, the working professional is probably equally well described by fuzzy-trace theory and the other dual-system or dual-process theories. Intuition is still an advanced form of thinking, more characteristic of experts rather than novices (in line with FTT), but at the same time applies standard procedures demanding not too much mental effort—a lot like the relatively artless protagonist, the intuitive System 1, in Kahneman/Epstein's paradigm.

4.2 Explaining Cognitive Phenomena

What comes before the analytical means just discussed when one must analyze data? Apparently the answer is—one's own cognitive skills. In this section, we take a bird's-eye view on what science has discovered about those human abilities relevant for our topic.

In the middle of the 20th century, Jean Piaget (Piaget and Inhelder 1951) established that in their early teens, children already intuitively understand and apply the law of large numbers. Today we know that about that age, they have just overcome the class-inclusion illusions (Brainerd and Reyna 1990), i.e. they no longer fail to distinguish between subsets and supersets in simple cognitive tasks. The famous example here is this: When children are shown pictures of seven cows and three horses and are asked "Are there more cows than more animals?" up to 10-year olds typically answer "Cows". Soon they leave such fallacies behind them.

Grown-ups make other mistakes. Even when the problems they face are very clear, they distort probabilities in judgements with economic and other consequences. Generally, they mishandle risk and uncertainty. They often misconceive quantitative information in various ways, extensively documented in the scientific literature. They are unduly overconfident, especially when the work on hand is remotely related to an area in which they have expertise. The choice of words presenting a case is of crucial importance for how facts and figures are appreciated, judged, and used in decision-making. But what are the state-of-the-art scientific approaches to explain all of these findings?

The tradition since Nicholas and Daniel Bernoulli has been to formulate a handful of principles, based on common sense. We recall Nicholas Bernoulli's conclusion about what "any fairly reasonable man" in the St. Petersburg game should do, and how his nephew resolved the paradox by postulating the existence of subjective utility. Over the turbulent decades and centuries that followed, this search for first principles continued unabated and brought to fashion opposite truisms in alternating sequence. To Gerd Gigerenzer (1994) we owe an eloquent account about how the concepts of frequentist probability and subjective belief evolved from a state when both were in unison, then went through a "divorce" in the 19th century, to get close again on a couple of occasions—though under new names—in the 20th century. Gigerenzer put these developments in a historic context and it is disturbing to read how events like the French Revolution could shake scientists' faith in the reasonable person. Science, it turned out, was susceptible to the general social climate of its time to an unusually great extent.

The 1970s heuristics and biases approach of Kahneman and Tversky represented another serious effort to develop a new set of explanatory principles. The human mind was now considered to be judging probabilistic information by using a fundamental Prototype heuristic, and its most famous daughter, the Representativeness heuristic (Tversky and Kahneman 1974; Kahneman 2003). Ingenious experiments helped to understand and characterize these mechanisms as vital, but at the same time vulnerable and prone to systematic errors.

Soon, however, some of these ideas were shown to be problematic by Gigerenzer, who proved experimentally that all the famous cognitive illusions disappear when data are presented case-by-case, allegedly in the way all species have learned to accumulate statistical information during millions of years of evolution. In turn, new experimental evidence suggested that even Gigerenzer's theoretical framework was not accurate enough. His findings were reinterpreted by Reyna and Brainerd (2008), who theorized that human performance improved in probabilistic tasks with natural (case-by-case) representation due to a different and deeper cognitive mechanism—the disentangling of the subset-superset relations. Indeed, examples involving very low-probability events were found, whereby the statistical one-by-one format of data actually degrades human performance (Koelher and Macchi 2004).

In our time there seems to be a consensus that when people reason about probabilistic or uncertain information they do not resort to formal logic, but are carried away by compelling coherent stories that easily confirm or at least relate to existing stereotypes. In addition, causal relationships are "discovered" where none exist—a malaise to which even some scientists are not immune, especially those doing regression analysis too much, and simply paying lip service to what Kendal and Stuart established long ago:

> [...] statistical relationship, however strong and suggestive, can never establish a causal connexion; our ideas on causation must come from outside statistics. (Kendall and Stuart 1967).

Amid the powerful stream of research exposing human cognition's true limitations, Nisbett et al. (1983) suggested a somewhat optimistic view about our ability to deal with probabilistic information. They pointed out that at least since the 17th century, scientists have been applying statistical thinking to an enlarging set of problems, and with a lag in time, ordinary people have been following them. Moreover, it is common today for almost everybody to discuss weather forecasts, sporting bets, medical issues, demographic and other social problems in at least the simplest probabilistic and statistical terms. The individual's cognitive abilities may be constant, but Western society seems to be adding new improvements to its own capacity to grasp empirical information.

Yet, returning to the question how science deals with human cognition, we can draw a dissatisfying conclusion. The approach tacitly adopted has been one of suggesting new postulates that more accurately account for the currently established findings, only to undermine and revise them in the light of new experimental results. This development is reminiscent of how economic theories deal with the agent motivation. A more puzzling issue is that by inventing various conceptual structures like: experiential/intuitive system, two-system cognition, dual processes, statistical heuristics, non-statistical (prototype) heuristics, subset-superset disentangling cognitive mechanisms etc., we are in fact populating science with phenomenological theories of little explanatory value, in short—with homunculi. All too aware that these are crutches to be dispensed with, we get used to them and begin to like them. We even use the modern fMRI and other scanning technologies to map them on the brain, tending to forget that what we need is something else and something more: a *causal* understanding of what is happening in it.

4.3 Quantifying Everyday Intuitions

Because of its enticing simplicity and sufficient generality, one principle seems to be surviving and weathering the storms against it, at least in some of the social sciences. This is the idea that in our daily activities we draw conclusions based on previous experience. Mathematically speaking, it is called the Bayesian approach. It posits a way to estimate probability of a future event by using previous probabilistic information.

Philosophically, this method for producing knowledge is problematic and as such has been lambasted already centuries ago. In 1889, the French mathematician Joseph Bertrand summarized all criticisms in a sarcastic phrase, saying that we know that the sun will rise tomorrow because of "the discovery of astronomical laws and not by renewed success in the game of chance". Social phenomena, however, are not as imminent as the sunrise, and their causes are not as deterministic. There, the objective and the subjective are not divided by a clear-cut border, but coexist in a no man's land. Reality is not the facts—reality is the interpretation of the facts, declares the human resources managers' jargon.

Yet, cognition has its own laws, and often they can approach the precision of the laws in the natural sciences. This suggestion is not as erratic as it may seem. Cognitive scientists Griffiths and Tenenbaum (2006) made an important step in clarifying how exactly the human mind is good at experience-based quantitative assessments in daily life. Not surprisingly, it turned out that we are very good Bayesian statisticians. These scientists gave an adequate mathematical formulation to the problem, which is described in this section.

The first tenet in their study was the assumption that people are capable of grasping the probabilistic distribution of events that are frequently repeated in their routine activities. Various natural and social phenomena have very different distributions, but the mind—the postulate states—is good at (unconsciously) learning them all. Then comes the second step. Not only the distribution shape becomes part of our tacit knowledge, but so do its essential statistical characteristics. For example, seeing a 30-, 40-, or 60-year old person, we assume that he/she will live much longer, while our expectations for a 95-year old are not that optimistic. To take another example, meeting a member of parliament in his 15th year in office may prompt us to ask ourselves how long will that person's political career last altogether. Probably not much longer either, but in a different way. While life expectancy is well described by the Gaussian distribution, political careers are best fit by the Erlang distribution. Various other matters have their own empirical distributions—such as, for example the time to bake cakes—that are not easy to

Table 4.1 Problems from everyday phenomena

Imagine you hear about a movie that has taken in 10 million dollars at the box office, but don't know how long it has been running. What would you predict for the total amount of box office intake for that movie?
If you heard a member of the House of Representatives had served for 15 years, what would you predict his total term in the House would be?
Imagine you are in somebody's kitchen and notice that a cake is in the oven. The timer shows that it has been baking for 35 min. What would you predict for the total amount of time the cake needs to bake?
...

Source Griffiths and Tenenbaum (2006)

describe with a simple formula. Griffiths and Tenenbaum discovered that people are generally able to give precise estimates in most such occasions.

Table 4.1 shows a number of examples from their study. People have been given the task to estimate how much longer a period will last, or how much more a quantity will rise, if they know the current amount, as stated in the particular situation. What all the problems described in Table 4.1 have in common is the demand to forecast quantity T, which may be the duration of a process or the total amount of some stuff.

Let us assume that we find ourselves in moment t, which is sometime between the beginning and the end of the process, i.e. $0 \leq t < T$. Let there be a set of subjective forecasts for its entire duration T. We take their median T^*. If the pool of people is large enough and each person is by nature a Bayesian statistician, then the subjectively assessed median T^* should be close to the median of the empirical distribution. In short, the Griffiths–Tenenbaum experiment sought to establish how close people's judgements were to the optimal statistical estimates in each task.

Obviously, this assessment of median T^* is an example for an operational definition because it has a strong link to measurement, as discussed in Chap. 1. The median can also be defined theoretically in terms of probability theory. The essential question here is how good the agreement between the theoretical and the operational definition is in the case of human judgement. I will say in advance that Griffiths and Tenenbaum discovered a remarkably good match in general, which is their great achievement. When the agreement was less accurate, it was for very interesting reasons, and those cases revealed important new facts about human cognition.

The theoretical median T^* can be determined from the classical definition of the concept in Bayesian context. We can write:

$$p(T > T^* | t) = \frac{1}{2}. \tag{4.8}$$

According to Eq. (4.8), T^* is the point for which there is a 50 % probability that the true value of T is greater than T^* and a 50 % probability that the true value of T is less than T^*. Our purpose is to find an analytical expression for T^* and then to compare it with the empirically found median as already discussed. To this end, we compute the posterior distribution over T given we are currently in t, or in short, $p(T|t)$. This is achieved by using the Bayesian formula:

$$p(T|t) = \frac{p(t|T)p(T)}{p(t)},$$ (4.9)

where

$$p(t) = \int\limits_{0}^{\infty} p(t|T)p(T)dT.$$ (4.10)

Let us clarify Eq. (4.9), illustrating it with the example of political careers. Quantity $p(T|t)$ defines the distribution of career durations in general. In particular, this may be the stint of a member of parliament (MP) for somebody who has been occupying the post for t years already. Quantity $p(t|T)$ is the probability for a particular MP with whom we are talking right now to have been in parliament for t years if MPs' careers last T years. By common sense, we assume that we could meet an MP at any point in their career with equal probability. This implies that t is a uniform-distributed random variable and for $T \geq t$ it follows

$$p(t|T) = \frac{1}{T}.$$ (4.11)

Apparently, for $T < t$ is true that $p(t|T) = 0$.

Now, we can take the *rhs* of (4.11) and substitute for $p(t|T)$ in the integrand of Eq. (4.10). Paying attention to the integration interval, we get

$$p(t) = \int\limits_{t}^{\infty} \frac{1}{T}p(T)dT.$$ (4.12)

Equation (4.12) is an important result. It states that for the class of problems discussed here, the distribution $p(t)$ for any given t is determined entirely by the prior distribution $p(T)$. Now let us continue with the example of political careers. The Erlang distribution—the one describing them best—in a relevant one-parameter special case has this form:

$$p(T) = T \exp\left(-\frac{T}{\beta}\right),$$ (4.13)

where β is a parameter. We substitute with the *rhs* of (4.13) into Eq. (4.12) and obtain:

$$p(t) = \int_t^\infty \frac{1}{T} T \exp\left(-\frac{T}{\beta}\right) dT = -\beta \exp\left(-\frac{T}{\beta}\right)\Big|_t^\infty = \beta \exp\left(-\frac{t}{\beta}\right). \quad (4.14)$$

Now we can substitute the expressions from Eqs. (4.11), (4.13), and (4.14) for all the quantities in the Bayes' formula (4.9):

$$p(T|t) = \frac{\frac{1}{T} T \exp\left(-\frac{T}{\beta}\right)}{\beta \exp\left(-\frac{t}{\beta}\right)} = \frac{1}{\beta} \exp\left(-\frac{T-t}{\beta}\right), \quad (4.15)$$

where $T \geq t$. Equation (4.15) is the analytical form of the sought posterior distribution given the Erlang prior. To compute T^*, we must take into account the following definition-like fact:

$$p(T > T^*|t) = \int_{T^*}^\infty p(T|t) dT. \quad (4.16)$$

Recalling (4.8), we set the *rhs* of (4.16) to be equal to ½. We then substitute what we found in (4.15) for $p(T|t)$ in Eq. (4.16). Thus, we obtain:

$$p(T > T^*|t) = \int_{T^*}^\infty \frac{1}{\beta} \exp\left(-\frac{T-t}{\beta}\right) dT = -\exp\left(-\frac{T-t}{\beta}\right)\Big|_{T^*}^\infty$$
$$= \exp\left(-\frac{T^*-t}{\beta}\right). \quad (4.17)$$

Combining Eqs. (4.17) and (4.8), we have

$$\exp\left(-\frac{T^*-t}{\beta}\right) = \frac{1}{2},$$

which implies

$$T^* = t + \beta \ln 2. \quad (4.18)$$

Equation (4.18) finally gives the analytical form of the posterior median for an Erlang-distributed variable. It depends only on the parameter β whose value is determined by the empirical data. Now this theoretically defined median can be computed and compared with the median of people's subjective judgements. Figure 4.3 shows the result.

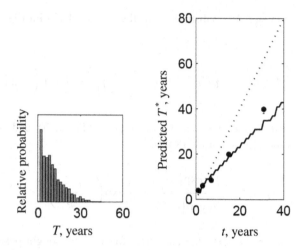

Fig. 4.3 *Left* Empirical distribution of the duration of US representatives' stints (2,150 representatives in the period 1945–2003). *Right* Optimal Bayesian prediction based on the empirical distribution (*solid line*) and participants' median predictions of T (*black dots*). Error bars indicate 68 % confidence intervals (Adapted from Griffiths and Tenenbaum 2006, with the permission of Sage Publications)

The five black dots in the right plot of Fig. 4.3 correspond to different values of t, varied randomly among the participants in the experiment. Apparently, the match between theoretical estimate (solid line) and empirical measurement (black dots) is extremely good. This means two things at once: people are able to recognize the exact kind of probability distribution (Erlang) and to assess correctly its concrete realization by giving answers that imply correct β. It must be stressed that these achievements are in no way related to formal statistical training and demand no knowledge of probability theory, parameter estimation etc.—we use such concepts only as a language to describe natural human abilities. In addition, the participants in the experiments were US undergraduates from whom no knowledge of history or political theory was expected.

Coming back to the right plot of Fig. 4.3, only the right-most black dot looks substantially misplaced. It corresponds to $t = 31$, which suggests that people's judgement becomes less precise when rare events are in the focus of attention. After all, 30–40 years US Representative careers are the exception rather than the rule.

The same undergraduates performed just as marvelously in a number of other tasks, very different but all related to daily phenomena. Figure 4.4 shows three such examples. Its left column shows the empirical distribution of life expectancy, which is well described by the Gaussian curve when infant mortality is excluded.

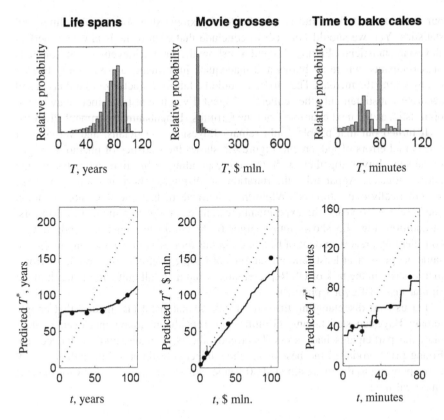

Fig. 4.4 *Top row* Empirical distributions of people's life expectancy according to US actuarial tables (*left*); total income of movies, 5302 in number (*middle*); and time to bake cakes, 619 recipes (*right*). *Bottom row* Optimal Bayesian prediction based on the empirical distribution (*solid line*) and the participants' median predictions of *T* (*black dots*). Error bars indicate 68 % confidence intervals [Adapted from Griffiths and Tenenbaum (2006), with the permission of Sage Publications]

The median of the pool of participants' predictions is accurate for all age groups. The middle plot shows an example of a power law distribution of the type

$$p(T) = T^{-\gamma},$$

where γ is a parameter. Again, with the exception of the blockbuster movies, the undergraduates' median estimate is quite correct.

Finally, the empirical distribution of the time needed to bake various cakes has a shape, which is quite complicated and not too easy to describe analytically. On the other hand, food is a topic much more important than many others are, and it is not surprising that people's medians have been very precise, as seen from the right column in Fig. 4.4.

These and other examples show people's cognitive ability to recognize and assess various probability distributions in daily matters. This is a performance of our intuition at its best and is not affected by knowledge about probabilities and statistics. Yet, we should not rush to conclude that each of us is always a perfect Bayesian statistician. Figures 4.3 and 4.4 show the joint achievements of hundreds of participants, where all biases and inadequate judgements are efficiently filtered out by taking the median. Therefore, it would be fair to characterize as unbiased and accurate statistician only the "collective" agent. Even that last statement may be too optimistic, as the next example from the Griffiths–Tenenbaum experiment suggests.

It turned out that people could remain Bayesian statisticians, and yet commit substantial errors of judgement. Figure 4.5 shows the statistics of pharaohs' reigns in ancient Egypt together with the corresponding subjective estimates of the undergraduates. Apparently, the duration of all reigns, short and long, has been systematically overestimated. While the empirical median over the entire pharaoh database was 16 years, the experimental estimates produced a median of 30 years. In addition, Fig. 4.5 shows much larger 68 % confidence intervals, indicating a considerably greater amount of hesitation in the appraisers. It is intriguing to see the same people—or at least random samples of people from the same pool—perform marvelously in the task of US Representatives and then fall quite off the mark in the similar task of Egypt's pharaohs.

Let us start disentangling the issue from the statement just made—that people remain Bayesian statisticians. Griffiths and Tenenbaum found that even in their bias, the participants had assessed correctly the kind of distribution. Indeed, the Erlang prior produced the best fit for the subjective estimates, but with $\beta = 17.9$ corresponding to the overestimated 30 years, rather than with $\beta = 9.34$ for the true empirical median.

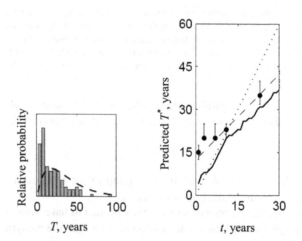

Fig. 4.5 *Left* Empirical distribution of the duration of pharaohs' reigns in ancient Egypt (126 pharaohs). *Right* Optimal Bayesian prediction based on the empirical distribution (*solid line*) and participants' median predictions of T (*black dots*). *Dashed line* shows a Bayesian prediction based on a subjectively overvalued prior. Error bars indicate 68 % confidence intervals [Adapted from Griffiths and Tenenbaum (2006), with the permission of Sage Publications]

Naturally, the typical respondent could not be expected to know much about life in ancient Egypt. The idea that life span was shorter then must have been close to their mind but apparently had little impact. This is perhaps a typical example for the well-known Anchoring heuristic—an unknown quantity is assessed in two steps, whereby an initial estimate (life longevity and monarch stint in our time) serves as a first iteration, to be made more precise in the second step. Kahneman and Tversky's studies have established that the subsequent adjustment is generally insufficient.

It is possible that a mix of two other cognitive mechanisms—Gigerenzer's Recognition heuristic, and Kahneman and Tversky's Availability heuristic—are additionally responsible for the unrealistic estimate. One can speculate that the typical respondent—an undergraduate from an elite US university—knew some history of the English-speaking world and was vaguely aware of the 63-year reign of Queen Victoria and the 50 years of Queen Elizabeth II on the UK throne by the time of the survey. That hypothetical respondent, however, probably had little further knowledge in royal matters, and it was these two examples that mostly informed their orientation about how long a monarch would typically rule.

Thus, it becomes clear how people could accurately identify the distribution shape and at the same time produce such an aberrant median. Griffiths and Tenenbaum (2006) provide a compelling summary:

> Given an unfamiliar prediction task, people might be able to identify the appropriate form of the distribution by making an analogy to more familiar phenomena in the same broad class, even if they do not have sufficient direct experience to set the parameters of that distribution accurately. [...] Such a strategy of prediction by analogy could be an adaptive way of making judgements that would otherwise lie beyond people's limited base of knowledge and experience.

So, when should the individual have faith in their ability for intuitive quantification? Mostly, in two domains. The first is her or his routine activities, in which practice including many trials and errors has established the optimal course of action, while emotion has largely given way to habit. The second domain is our professional activity—there again we have gathered experience, which has been amalgamated with education and training, designed to equip us with answers to the set of typical cases we are likely to come across. However, it takes only a step or two outside of these spheres, for life to show us how susceptible to framing by verbal formulations and cognitive illusions we are, and how far off the mark our judgement can imperceptibly become.

References

Brainerd, C. J., & Reyna, V. F. (1990). Inclusion illusions: Fuzzy-trace theory and perceptual salience effects in cognitive development. *Developmental Review, 10*, 365–403.

Gigerenzer, G. (1994). Why the distinction between single-event probabilities and frequencies is important for psychology (and vice versa). In: G. Wright and P. Ayton (Eds.), *Subjective Probability*, (pp. 129–161). Hoboken, NJ: John Wiley.

Part III
Intuition and Decisions

Chapter 5
The Decision Maker's Neural Apparatus

5.1 A Need for a New Scientific Method

In the preceding chapters, we covered two major approaches to decision analysis—the axiomatic, associated mainly with expected utility theory, and the psychological. Here I outline a third approach, which is the *neurobiological modelling of the decision maker*. It differs greatly from the former two as it employs an entirely new set of concepts, postulates, and mathematics. In its roots, there lie three sciences: biophysics, neurophysiology, and psychology. Initially this field was not connected with decision-making, let alone with economics or management science. Its primary objective was to formulate axioms guiding the study of the structure and functioning of the human brain. This road was intended to eventually produce a description, similar to a computer's technical specification. During the 1940s and 1950s there has been little understanding how daunting these problems were, and such a goal seemed within reach in the near future.

Then, and even today, computer scientists have been enticed by the idea to develop Artificial Intelligence (AI)—a machine able to replace its creator in places where it is impossible, unacceptable, unjustifiable or simply undesirable to have humans operating. AI could probably solve complex large-scale optimization and control tasks in real time. The demand that a machine be capable of high quality decision-making naturally provoked an interest in the same process in humans. However, the latter are tricky: they do not quite think rationally but rather employ cognitive-emotional interactions, and these have for long been too difficult to grasp and implement in artificial intelligence.

One bright idea has been to try to understand how the brain works at a micro level, where neurons interact. In the present chapter, I put this level of analysis in focus and briefly follow its development over time. The neurobiological approach might become especially important in the future because its potential success would improve our understanding of human decision-making in general. Indeed,

© Springer-Verlag Berlin Heidelberg 2015
G. Mengov, *Decision Science: A Human-Oriented Perspective*,
Intelligent Systems Reference Library 89, DOI 10.1007/978-3-662-47122-7_5

developing adequate postulates directing the studies of cognitive mechanisms would greatly help overcome the problematic state of today's decision science.

In Chap. 3, the discussion of prospect theory ended by mentioning how it could not account for at least 11 paradoxes of choice behaviour discovered after its publication, but an alternative theory, TAX, (Birnbaum 1999, 2008) was believed to be able to explain them. With the established tradition in mind, one can expect that its validity claims would soon be challenged by new instances of paradoxical behaviour, if only TAX managed to attract sufficient attention. Apparently, such state of affairs is hardly satisfactory. However complex the cognitive mechanisms employed by humans in their decision acts might be, it is unacceptable that the axiomatic base of decision science should be reformulated in response to each new psychological discovery.

What is needed, therefore, is an entirely new approach able to offer, in von Neumann and Morgenstern's (1944, 1.3.) words, "...*methods... which could be extended further and further.*" Put differently, decision science needs a new paradigm—a novel system of attitudes, values, and techniques adopted by the research community (Kuhn 1962, 1970). It should,

> ...in the future guide research on problems many of which neither competitor can yet claim to resolve completely. A decision between alternate ways of practicing science is called for. The decision is based on future promise rather than on past achievement. (Kuhn 1962, Chap. XII.)

A new paradigm would initially appear unconvincing because not only it has not had the opportunity to show its merits, but also because its initial formulation is always far from perfect; moreover, a radically new scientific approach is usually difficult to understand. Its creators and pioneers are in a delicate situation, and sometimes might benefit from the advice to approach a field ..."*with a sense of modesty...*" (Neumann and Morgenstern 1944, 1.3. and 1.4.) Yet, the founders of game theory have also suggested criteria to assess the maturity of any new theory: after an initial stage dealing with elementary problems, it would approach issues "beyond the obvious", and eventually reach the level of "genuine prediction".

Neurobiological modelling of the decision maker is one such new paradigm. It has existed for many decades and has already reached adulthood, allowing predictions exceeding that of some competing methods. The key players in this field are neurons, synapses, and neurotransmitters. Their activities and interactions are embedded in models of growing complexity, which eventually reach the macro level of observable human behaviour.

This chapter is interdisciplinary in character and might be too complicated for some readers. To reach out to them, in Sects. 5.2 and 5.3 I explain the essentials of the neurobiological method in an approachable manner, omitting all mathematical formulae. Then, the exposition continues with the description of cognitive processes with systems of ordinary nonlinear differential equations.

5.2 Neurobiology for Modelling Cognition

It is accepted that this branch of science started in 1943 with an article by McCulloch and Pitts (1943) in *Bulletin of Mathematical Biophysics*, followed in 1947 by McCulloch and Pitts (1947) in the same journal. There they used propositional logic and introduced a system of axioms describing most of what was known about the functioning of the nervous system at the time. Their postulates are summarized in Box 5.1. These ideas became very influential for decades to come.

Box 5.1 McCulloch and Pitts postulates for the nervous activity
By the 1940s a number of properties of the neurons and their interactions were already discovered. McCulloch and Pitts (1943) summarized the existing knowledge in the following set of postulates, called by them *physical assumptions*:

1. The activity of the neuron is an "all-or-none" process.
2. A certain fixed number of synapses must be excited within the period of latent addition in order to excite a neuron at any time, and this number is independent of previous activity and position on the neuron.
3. The only significant delay within the nervous system is synaptic delay.
4. The activity of any inhibitory synapse absolutely prevents excitation of the neuron at that time.
5. The structure of the net does not change with time.

Here the neuron was treated as a logical device turned on and off under certain conditions. This simplicity fascinated some researchers who suggested that there existed analogies between the neuron and the building components of the electronic computers of their time. Human brain began to be seen as a better kind of a computing machine. At about the same time a reciprocal idea was born—to design more advanced computers, eventually developing Artificial Intelligence, by emulating the brain and its functions.

During the 1950s it became clear that the living matter, even in its simpler forms, was a lot more complex than the pioneers of early computer science imagined. A milestone was set by a 1959 scientific article entitled "What the frog's eye tells the frog's brain" and authored by Lettvin, Maturana, and the same McCulloch and Pitts. Studying the anatomy of connections between the retina and the inner neural layers in this species, they discovered four interconnected groups of fibers serving four different functions of image processing. An important research conclusion was that,

> ... the eye speaks to the brain in a language already highly organized and interpreted, instead of transmitting more or less accurate copy of the distribution of light on the receptors. (Lettvin et al. 1959).

Offering an example of scientific integrity, these authors admitted that,

> ... The operations found in the frog make unlikely later processes in his system of the sort
> described by two of us earlier (McCulloch and Pitts 1947).

This setback attracted attention to the need for new modelling approaches in
neurobiology and eventually led to the birth of neural networks. This scientific area
benefited immensely from a boost by Stephen Grossberg, a psychologist with a
doctorate in mathematics. His early contributions date from the 1960s, yet he
continued to do important work even in the 21st century. Like McCulloch and Pitts,
Grossberg investigated how a small number of neurons work to exchange signals.
Unlike his predecessors, however, he always started the analysis from 'above', i.e.
from observed behaviour—psychological reactions and actions in humans and other
species. To this end, he first studied scores of scientific articles with biological and
psychological experiments dealing with a particular cognitive phenomenon. In the
multitude of empirical data, he sought to discover what would emerge as typical
and common, and used it to formulate new postulates about the organization and
functioning of the nervous system.[1]

The next step would be to develop a mathematical model embedding the new
postulates. It must be the simplest among a set of candidates in terms of elements
and relations, yet should adequately describe the particular mechanism of interest.
At this stage, computer simulations become relevant and help quantify aspects of
the studied phenomenon. Independently, the model's formal mathematical properties are investigated and necessary theorems are proven. Eventually, a model
emerges which is most parsimonious (satisfying Occam's razor) and at the same
time best predicting, within the limitations of the pre-defined set of principles.

At the next stage, one can analyze the variations in anatomy and behaviour that
different species exhibit with respect to the phenomenon, and how variants of the
basic model or its parameter selection can account for them. Going this way, the
model is further applied to cases of greater complexity until its predictive power
begins to deteriorate as it reaches its boundaries of validity. There, the known meets
the unknown and it becomes clear which theoretical simplifications are no longer
justified and must be abandoned. New postulates are formulated to join the existing
and serve as guidelines for developing more sophisticated models of complicated
behaviours.

The need to embed simpler in complex models has provoked an unexpected
challenge. It has turned out that the act of embedding becomes a reality check for
the set of postulates on which the simple models had rested. Only postulates and
models which are able to interact adequately in the bigger neural structure, would
eventually earn a place in a more powerful theory.

---

[1]Here lies the most creative act not only in this particular research, but also in science in general. It
is called the *inductive leap*, sometimes also referred to as *speculative leap*—the twist of thought in
which, due to available information, a new statement of explanatory value is generated. It was the
German poet Goethe, who noted that the highest accomplishment in science as well as in life was
to take a difficult problem and formulate a postulate out of it.

Grossberg's main achievement is that he successfully charted the road from simple neural interactions at micro level, to complex phenomena such as short-term memory, long-term memory, recognition of visual images and auditory signals, emotion generation, to end up with macro level behaviour such as reflex conditioning and cognitive-emotional interactions. For contemporary reviews, see (Grossberg 2009, 2013). Philosophers of science would perhaps define his method as a mixture of induction and a hypothetic-deductive (also called "Received View") approach. This new apparatus can produce models describing increasingly complicated phenomena, and thus meets the criteria formulated by von Neumann and Morgenstern, and Thomas Kuhn for a new paradigm with superior generalizing power.

Some of Grossberg's early publications (Grossberg 1967, 1969a) contain vocabulary about learning machines, modern in the 1960s; yet he says that the field of Artificial Intelligence never influenced him. His main interest was in the "natural intelligence", i.e. the successful adaptive behaviour and its relevant psychological mechanisms in humans and other species.

Because all brain processes take place in continuous time, it was natural to describe them with differential equations. By the middle of the 1950s it was known that a neuron's bioelectric potentials generate ionic currents in its longest projection—the axon, whereby proteins called neurotransmitters are released; as these float, they reach other neurons and change locally their biochemical properties. Three scientists: John Eccles, Alan Hodgkin, and Andrew Huxley were awarded the Nobel Prize in Physiology or Medicine in 1963 for their discoveries in the field. The latter two ingenuously made use of a differential equation describing electric signal in a cable, to explain the bioelectric signal moving along the axon of a neuron.

Grossberg (1969b) discovered analogies between some neurophysiological phenomena and his own learning machines. It became clear that two very different roads had led to essentially identical explanations of what is happening in the nervous system. In his subsequent work, Grossberg has used extensively the Hodgkin–Huxley equation (Hodgkin and Huxley 1952) alongside two other equations, introduced by him, to describe various aspects of behaviour.

One could observe that even to this day, virtually the same equations are employed by biophysics and mathematical psychology. In the former science, they are instrumental for computing bioelectric voltages, while the latter describes cognitive phenomena with no electric measurements, and that is why it uses them in dimensionless form. Apparently, this is an interdisciplinary area. In the early 21st century, it was called *computational neuroscience* or *mathematical neuroscience*, which signified a new stage beyond the traditional mathematical psychology.

Computational neuroscience developed rapidly in the last decades and its achievements are to be found in a variety of applications. Having started with the objective to study brain processing of visual and auditory information, speech generation, and motor control, this science now provides the basis for many technological implementations. Some of them include robotics, industrial and medical decision support systems, data base management, and satellite image recognition. Neuroscientific models are used in studying intelligent behaviour in humans and other species, and are embedded in artificial systems especially where learning, adapting to changing environments, and decision-making under risk or uncertainty are involved. New theories contribute to this picture every now and again (cf. Busemeyer and Townsend 1993; Litt et al. 2008; etc.)

5.3 From Neurons to Behaviour—A Brief Note Without Formulae

Grossberg based his theories on three fundamental mechanisms of neural activity and interaction: (1) Exciting a single neuron which makes it emit a signal to other neurons; (2) Transmission of a signal from one neuron to another via neurotransmitters; (3) Remembering a unit of information due to long-term biochemical change in a neuron receiving a signal. Each mechanism has been described with a nonlinear ordinary differential equation. Their joint work gives rise to complex neurophysiological processes, which on the large scale comprise cognitive phenomena. At that level, multitudes of components of the type (1)–(3) are present and each is modelled with an equation.

A prominent example is the way in which humans and higher animals handle expected and unexpected events, associated with potential benefit or danger. Let us discuss a simple social situation. Imagine that someone is expecting a positive event involving another person—for example, jointly attending an opera performance. At the right moment, the second person neither shows up nor sends any signals to explain their absence. Formally, no event has occurred and therefore the stimulus–reaction type models are not relevant, as they would predict that the person in waiting would do so for an indefinite period. Of course, nothing like that happens in real life. Imagine what could be happening in the head of the waiting person: First, a positive emotion of anticipation occurs, to be followed by disappointment due to the failed rendezvous. At the peak of anger, he or she takes a new decision—to leave the opera and do something entirely different. That new activity helps alleviate the negative emotion, which gradually abates and opens the way for new experiences. This altering of opposite emotions is known as *opponent processing*— a vital mechanism for adaptive behaviour, as it provides the drive to take adequate actions in response to the flow of environmental challenges.

Grossberg has formulated a mathematical model to account for opponent processing and has devised a neural network, which is its realistic biological implementation. This is the Gated Dipole neural network (Grossberg 1972). The term *dipole* denotes a structure in which two opposite entities, in this case—psychic impulses, interact. In its simplest form it involves mechanisms of the type (1) and (2) to account for the consecutive generation of opposite emotions like fear and relief, joy and sorrow, hunger and satiation, satisfaction and disappointment, etc. A group of neurons in two information channels produces the main neurobiological effect here: one channel serves to generate a positive emotion due to a beneficial external event; the other generates a negative emotion in response to a harmful event. The two channels are connected in such a way that they continuously try to suppress each other, thus engaging in a dynamic balance. Their interaction gives rise to a sequence of states determining the individual's dominant emotion at any moment. The substrate of this affective dynamics is the quantity of neurotransmitters, accumulated and released in each of the two channels.

The gated dipole is able to account for a variety of reactions exhibited by people in their social interactions. It could be argued that dipole action is at the heart of behaviours such as neurotic selling and buying in the stock markets, a student's anxiety during an examination, or any other daily event. The gated dipole has been called for to explain not only emotions, but also has been utilized in models of image recognition, retino-cortical dynamics, motor control, and decision-making.

5.4 Analytical Account of the Neurobiological Method

5.4.1 Three Micro Mechanisms

Let us come back to the three neural mechanisms, assisted this time by some mathematics. To reiterate, those are:

1. Single neuron activation;
2. Signal transmission from one neuron to another via a synaptic connection;
3. Storing information in the memory by long-term biochemical change in a receiving neuron.

Starting with *Single neuron activation*, consider a nerve cell in a population of N neurons. The i-th one ($i = 1, \ldots, N$) sends signal y_i to other neurons, and this is its bioelectric activity, also called action potential. It is described by the classical Hodgkin–Huxley equation, cited below in a form typically used in computational neuroscience:

$$\frac{dy_i}{dt} = -a_1 y_i + (a_2 - y_i)J_i^+ - (y_i + a_3)J_i^- . \tag{5.1}$$

Here J_i^+ is the sum of all incoming signals from other neurons connected with neuron i and trying to activate it. Similarly, J_i^- sums all inhibitory signals from other neurons; a_1, a_2, a_3 are real and positive constants ($a_2 \gg y_i$). Hypothetically, should all incoming signals to i cease for a moment ($J_i^+ = J_i^- = 0$), then y_i would decay to zero at a rate, accounted for by the constant a_1. Equation (5.1) is widely used and all neuron equations in this book are its special cases. Because we are interested in the neuron in no other way but with regard to its activity, henceforth I will use "neuron", "neural activity", "activation", "neural signal", and "Short-term Memory (STM)" as synonyms and will denote them all by y with a subscript. In addition, all equations will be in dimensionless form.

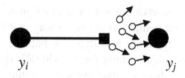

Fig. 5.1 Neuron i sends signal to neuron j by emitting neurotransmitters

Signal transmission from one neuron to another via a synaptic connection is the second mechanism (Fig. 5.1). A neuron sends signal by emitting neurotransmitters, also called mediators, which are protein molecules floating in the synapse—the area where two neurons' projections come close to each other. These neurotransmitters reach the receiving neuron and cause biochemical changes in it, altering its sodium ion and potassium ion concentrations and thus creating (bioelectric) action potentials in it.

Under the assumption that a signal must not be too much distorted when transmitted, it is desirable that the receiving neuron's activity be proportional to the sending neuron's activity throughout the entire duration of a signal. Yet, due to exhaustion of the neurotransmitter quantity z_i in the sending neuron, and the needed time to replenish it, such a demand is unrealistic. Therefore, mediator release and regeneration is a nonlinear process in time, as described by Eq. (5.2), Grossberg (1972, 1998):

$$\frac{dz_i}{dt} = b(1 - z_i) - cy_i z_i. \tag{5.2}$$

Here b and c are positive constants. Derivative dz_i/dt gives the rate of mediator loss and replenishment in the sending neuron i. Term $b(1 - z_i)$ accounts for transmitter regeneration. It is produced until the maximum capacity (equal to 1) is reached. The regeneration rate is accounted for by the constant b. Term $cy_i z_i$ states that mediator loss in the sending neuron is proportionate to $y_i z_i$. That is, the stronger the sending signal y_i, the faster would neuron i empty its store. But reproduction takes place simultaneously, and so a kind of dynamic balance is achieved.

Let us for simplicity consider an instantaneously changing signal y_i, which is then kept constant for some time. This is an idealization helping us solve analytically Eq. (5.2). Indeed, with $y_i = const$, its solution is this:

$$z_i(t) = \frac{b}{b + cy_i} + c_I \exp[-t(cy_i + b)]. \tag{5.3}$$

With c_I being the integration constant, the equilibrium value for z_i is given by $z_i(\infty) = b/(b + cy_i)$. Having in mind the actual behaviour of the receiving neuron as described by Eq. (5.1), the signal transmission via the synapse is illustrated in Fig. 5.2. Because this process is much slower than the neuron activation, it is called Medium-term Memory (MTM).

Fig. 5.2 Transmission of a
signal from neuron i to
neuron j. *Top* Idealized input
signal. *Middle* Change of
mediator quantity z_i in the
sending neuron. *Bottom*
Activation of the receiving
neuron j

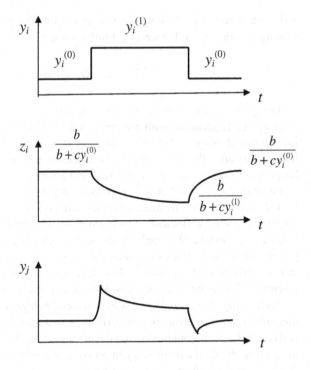

The last of the three mechanisms is *Storing information in the memory by long-term) biochemical change in a receiving neuron*. A number of facts are important here. First, what we call *memory*, are permanent biochemical changes in the connection between two neurons. Second, the new information must not destroy any other knowledge that a person had accumulated before. Therefore, synaptic alterations due to new signals must happen selectively and "locally"—only in places where the previous version of a particular chunk of knowledge is kept, or an entirely new knowledge is to be stored.

Third, the incoming stream of data is multidimensional and is transmitted to the brain by a vast number of neural cells acting in parallel—the receptors for light, sound, taste, tactility, and odour. This information can be represented by a huge matrix with elements changing continuously. For example, the human retina contains two kinds of receptors: six to eight million cones responsible for visual sharpness and color perception; and 120 million rods, providing orientation in twilight. Interpreting such data volumes needs sophisticated coordination among nerve cells that are involved, and excluding all the rest. Because of the continuous-time nature of these phenomena, once again it is appropriate to describe them with differential equations. There are many ways to do so, and historically the first, and still a convenient one, has been Grossberg's *Gated Steepest Descent Learning Law* (Grossberg 1969a, b, 1998; Grossberg and Schmajuk 1987). It is given by Eq. (5.4) and models how a single long-term memory component z_{ij} between neurons i and j changes under the influence of a third neuron k:

$$\frac{dz_{ij}}{dt} = y_i(-h_1 z_{ij} + h_2 y_j). \tag{5.4}$$

The gated learning law by Eq. (5.4) is different from Grossberg's early publications, but is identical with the one used by Grossberg and Schmajuk (1987). Long-term Memory (LTM) element $z_{ij} \in [0, 1]$ is a dimensionless variable with initial value often (but not always) assumed to be zero, meaning that no learning has happened. Constants h_1 and h_2 are real and positive.

Equation (5.4) says that z_{ij} can change only when neurons i and j are simultaneously active. A third neuron k provides activation for j. In particular, a nonzero y_i allows the memory element z_{ij} to change. In that sense, neuron i *gates* learning. Updating z_{ij} can take place only in moments when i is active ($y_i > 0$). When $y_i = 0$, previously learned values of z_{ij} stay the same. Neuron j provides the signal to be learned, influenced by neuron k (Fig. 5.3). The magnitude of signal y_j carries the essential information that turns into knowledge and is stored in z_{ij}.

Technically, there is only one way to remember signal y_j. According to Eq. (5.4), memory element z_{ij} changes until term $h_1 z_{ij}$ becomes equal to term $h_2 y_j$. Then the derivative dz_{ij}/dt becomes zero and the process stops. In other words, we remember in z_{ij} a tiny chunk of a momentary impression or emotion, carried by neural signal y_j that persists for a short while. Therefore, memory element z_{ij} remembers an averaged value of the signals that have been coming from neuron j, up to a scaling coefficient. As the memory unit tracks the time-average of the information signal, element z_{ij} can go up or down due to the latest update. This neural network can be

Fig. 5.3 Remembering a chunk of information—signal y_j, in the synapse z_{ij} between neurons i and j. The synapse is the smallest, or unitary, unit of information in the brain

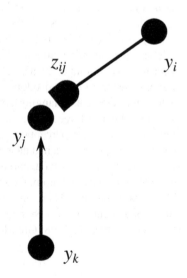

phenomenologically interpreted as performing something remotely similar to Bayesian learning (Grossberg and Pilly 2008).

5.4.2 The Decision Maker's Emotions

Let us now consider how the three elementary mechanisms work together to generate an emotion in response to an important event. Central in this process is the gated dipole, which models interactions among neurons. Its equations are the following:

$$\frac{dy_1}{dt} = -y_1 + I + J \tag{5.5}$$

$$\frac{dy_2}{dt} = -y_2 + I \tag{5.6}$$

$$\frac{dz_1}{dt} = b_1(1 - z_1) - c_1 y_1 z_1 \tag{5.7}$$

$$\frac{dz_2}{dt} = b_2(1 - z_2) - c_2 y_2 z_2 \tag{5.8}$$

$$\frac{dy_3}{dt} = -y_3 + y_1 z_1 \tag{5.9}$$

$$\frac{dy_4}{dt} = -y_4 + y_2 z_2 \tag{5.10}$$

$$\frac{dy_5}{dt} = -y_5 + (1 - y_5)y_3 - (y_5 + 1)y_4 \tag{5.11}$$

$$\frac{dy_6}{dt} = -y_6 + (1 - y_6)y_4 - (y_6 + 1)y_3 \tag{5.12}$$

$$o_1 = [y_5]^+ \tag{5.13}$$

$$o_2 = [y_6]^+. \tag{5.14}$$

All $y_1, \ldots, y_6$ are neural activations, z_1 and z_2 are neurotransmitters, o_1 and o_2 are the positive components of neural signals y_5 and y_6, and I, J are input signals. Constants b_1, b_2, c_1, c_2 are real and positive. A closer look at Eqs. (5.5), (5.6), (5.9)–(5.12) reveals that they are all special cases of the Hodgkin–Huxley Eq. (5.1). For example, the input signal y_3 in Eq. (5.11) plays the role of activating sum J^+, while y_4 corresponds to J^-. Equations (5.7) and (5.8) coincide with (5.2), describing neurotransmitter release. Both Eqs. (5.13) and (5.14) define the dipole's output signals comprising the opposite emotions. The notation $[.]^+$ signifies mathematical rectification, $[\xi]^+ = \max[\xi, 0]$. Here it is used to allow only one of the opposite emotions to exist at any one time; the intensity of the other must be equal to zero.

How this happens is shown in Fig. 5.4. Because it is not known precisely how gated dipoles match the real anatomy, it cannot be specified whether $y_1, \ldots, y_6$ represent single neurons or homogeneous groups of neurons. In both cases, however, the analysis yields identical results.

The system of Eqs. (5.5)–(5.14) contains six neurons and only two synapses, which is a simplification. Using it is justified because these elements are sufficient to clarify the dynamics of opposite emotions. Inclusion of more synapses would bring in nothing substantial, and therefore it is appropriate to apply Occam's razor.

Figure 5.4 contains another idealization. The actual signal transmission from y_1 to y_3 via synapse z_1 works exactly as it was shown in Fig. 5.2 where the outgoing signal is a lot more distorted. In contrast, all signal fronts in Fig. 5.4 are shown to be

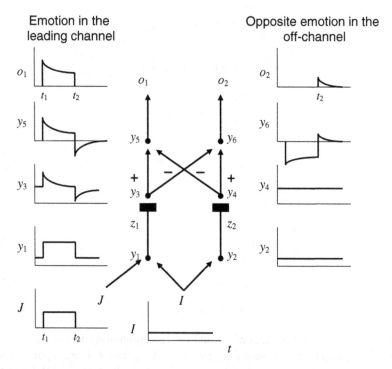

Fig. 5.4 Generating opposite emotions in a gated dipole neural network. For example, during an exam taking place between moments t_1 and t_2 a student feels stressed (output o_1, *upper left corner*). This emotion is intense at the beginning and then abates, but does not disappear until the very end, t_2. After the exam the student feels relief and perhaps even joy (output o_2, *upper right corner*) for some time. Plus and minus signs around neural pathways indicate activation and inhibition signals

vertical, as if they change instantaneously. In reality, they take seconds and even minutes to change, but these transient processes can be neglected because the states in-between last for hours. For example, a student exam may take a couple of hours, but its beginning and end happen within minutes.

Let us see how the dipole functions. First, each neuron receives a constant *tonic* signal whose role is to keep it in working condition. Only in such a state, a neuron is able to receive informative signals and transmit them. How important is this can be seen from the fact that a neuron at work consumes only 2–4 % more oxygen than a "resting" neuron. Back to the dipole, input neurons y_1 and y_2 receive as tonic signal I, as seen from Fig. 5.4 and Eqs. (5.5)–(5.6). It is assumed that all the other neurons are put in working condition by the constant components of their input signals, or simply Occam's razor is applied again.

A gated dipole contains two information channels with neurons y_1, y_3, y_5 and y_2, y_4, y_6 respectively. An external event such as the start of an exam for a student, or some business news for a trader, sends information signal to only one of the channels. This is signal J in Fig. 5.4, which appears at moment t_1 and disappears at moment t_2. For clarity, from now on we call the set of neurons y_1, y_3, y_5 *leading channel* as its function is to generate the currently dominant emotion. The other channel contains y_2, y_4, y_6 and will be called *opposing* because it produces the opposite emotion. In various applications, the leading channel can emit positive or negative emotion depending on the concrete circumstances. Both channels are linked in such a way (see Eqs. (5.11)–(5.12) and Fig. 5.4) as to compete and try to suppress each other's output signal.

When input J is submitted to y_1 it is added to I according to Eq. (5.5) and is then further transmitted to y_3 and y_5 exactly as was shown in Fig. 5.2. However, due to Eq. (5.11), neuron y_5 also receives an inhibitory signal from y_4 which reduces its total activity as compared to y_3. It is obvious from Fig. 5.4 that signal y_5 has the same shape as y_3 but is shifted downwards, with the difference being equal to y_4. Finally, as per Eq. (5.13), the emitted emotion o_1 is only the positive part of y_5.

Input J invests its energy in the leading channel and helps it suppress the opposing channel between moments t_1 and t_2. Thereby, neurotransmitter is released only in the left synapse (Fig. 5.4) and z_1 is partly exhausted, exactly as was shown in Fig. 5.2. When at moment t_2 signal J stops abruptly, the dipole is in a situation with spent z_1 and full z_2. For a short while this shift of balance gives advantage to the opposing channel, which emits its own signal o_2, the opposite emotion. In light of this interaction, o_2 is also called *rebound* signal. Soon z_1 is replenished and becomes equal to the slightly exhausted z_2. Again equilibrium is established and the dipole stops emitting signals.

Now the name of this neural network becomes obvious. It is "gated" by an input information signal: when one exists, the dipole is active; otherwise, it remains balanced and reticent. Science has yet to clarify the exact anatomy of gated dipoles in the brain, but it is already clear that they embody a fundamental mechanism with a variety of applications.

5.5 Emotional Balance and Economic Decisions

Our daily activity offers many opportunities to employ opponent processing as a mechanism for understanding human behaviour. The economy, for example, often becomes a source of strong passions and anxieties. In contrast with other areas of social activity, all emotions it stirs stem from two clear sources—financial profit and loss. Markets have devised the metaphorical images of bulls and bears as symbols for the agents' optimism and pessimism. Warren Buffet has described this game of opposites in vivid words:

> The market behaves as if it were a fellow named Mr. Market, a man with incurable emotional problems. At times he feels euphoric and can see only the favorable factors, while at other times he is depressed and can see nothing but trouble ahead for both the business and the world.

Wall Street traders are known for neurotic actions of a specific type: aggressive buying of stock after good news, followed by nervous selling of the same stock no matter what the following news would be. Such hesitations can be naturally explained in the light of gated dipole functionality, and more specifically, by Grossberg and Gutowski's (1987) affective balance theory, which is an implementation of Grossberg's method to the domain of decision-making. This theory regards choices as cognitive-emotional acts and helps clarify the role of contextual information in decision situations—a long-standing issue in Kahneman and Tversky's work. A gated dipole is at the centre of affective balance theory and accounts for the agent's deliberations in real time.

According to Grossberg and Gutowski (1987), each effect explained by prospect theory can also be understood in dipole framework. In particular, the dipole sheds light on a number of fundamental behavioural phenomena, of which here I summarize only two. These are:

1. Shift of reference point due to positive or negative framing of a case;
2. The gambler's fallacy.

The first effect is illustrated by the paradoxical choices made by people in the case of losing 600 jobs due to business restructuring (Chap. 3, Sect. 3.2.3). Participant reactions can be interpreted with the help of a gated dipole in the following way. The danger of laying off people provokes a negative affect in the corporate manager or privatization agency boss which is materialized by a substantial neurotransmitter release in the dipole channel for negative emotions. Because replenishment is much slower than release, for some time there exists a combination of exhausted neurotransmitter in the negative emotions channel and unspent neurotransmitter in the positive emotions channel. At precisely that moment the manager considers the offer to save one third of the jobs (and tacitly close down the remaining 2/3). It is obvious that the statement, "Save 200 jobs for sure" (A_{II} in Table 3.1) generates a strong impulse in the positive emotions channel, made even stronger by the momentary neurotransmitter imbalance as explained above. Naturally, A_{II} has caused a strong positive affect.

In the next moment, alternative B_{II} (save all 600 with p = 1/3, or lose them with p = 2/3) is assessed. Now both channels in the dipole are receiving signals, and it is difficult to tell which is the stronger. At the same time, both channels have spent some of their neurotransmitters. It is hardly surprising then that B_{II} causes much less enthusiasm. Under the natural assumption that the emotional reactions to alternatives A_{II} and B_{II} have been stored in the dipole's neural memory and can be easily compared, it becomes obvious why the riskless A_{II} (save 200 with certainty) would win the competition—it has been supported by a much stronger positive emotion. Exactly this kind of reaction was observed in the experiment we conducted (Mengov and Hristova 2004; See also Sect. 3.2.3, Table 3.1). There, the riskless alternative A_{II} was chosen by 60–87 % of the participants in the two experimental conditions.

One essential conclusion is that the gated dipole model, by virtue of its theoretical foundation, can offer causal explanations at neurobiological level. It is noteworthy that it enables understanding of a micro mechanism, responsible for what is observed at the macro level of decision-making.

The second example shows how a gated dipole can provide insight about the type of behaviour, known as *the gambler's fallacy*. Psychology has long ago established that when people suffer a loss, they become willing to take risks above their usual risk appetite. This is of particular relevance for business people and casino gamblers, as both occupations involve risky decisions on ongoing basis, some of which ending in failure. By extension, it is obvious that the risk someone is willing to accept after a *sequence* of negative events is greater than what is acceptable after a mixed sequence of gains and losses. A gated dipole explanation states that each event in a business or a casino provokes neurotransmitter release in the channel for negative emotions (in response to a loss), or in the channel for positive emotions (due to a gain). A mixed series of events leads to neurotransmitter release in both dipole channels, and hence the agent retains a relatively balanced risk attitude. If, however, the agent has had a sequence dominated by losses, his or her momentary emotional balance is "skewed".

Due to neurotransmitter exhaustion in the channel for negative emotions, the unpleasant feeling after each new loss is attenuated, and this effect "dulls the edge" of the emotion as a warning mechanism against further spending of resources—financial or any other. Should the agent—businessperson or gambler—continue to act "intuitively", they are bound to take further risks whose assessment would not rest on expert opinion (at least for the business professional), but only on temporary neurotransmitter imbalance. Now the role of the dipole mechanism with self-suppressing channels becomes apparent—it exists in order to quickly reinstate the emotional *equilibrium*. An important prerequisite for adequate actions in general, the latter is particularly relevant in decisions where all sorts of resources are spent.

5.6 Reflex Conditioning and Consumer Behaviour

People learn from experience and try to avoid actions that could cause negative affect. This ability is based on the propensity to link in the mind events that jointly reoccur. There exist innate mechanisms of associating unconditioned stimuli such as food and water with neutral stimuli such as lamps, bells, advertisements, etc. When an unconditioned and a neutral stimulus are submitted together in a certain repeated procedure, the neutral stimulus becomes conditioned, thus instating a conditioned reflex. This mechanism was discovered in the early 20th century by the Russian physiologist Ivan Petrovitch Pavlov during his experiments with dogs. Influenced by his findings, many European languages have subsequently adopted the metaphor "Pavlovian dog" to signify habitual behaviour even in humans.

This type of associative learning is scientifically termed *classical conditioning*. Another type is the *operant or instrumental conditioning*, which combines a mixture of stimuli causing both positive and negative emotions, aimed at provoking certain desired behaviour. The American scientists Edward Thorndike and Burrhus Skinner were the pioneers in this field. Science has established yet other forms of learning, all of them aimed at adaptation to the environment.

The gated dipole is insufficient to take care of such a cognitive function. However, it can be augmented with emotional memory. During the 1980s, Stephen Grossberg together with Nestor Schmajuk and Daniel Levine introduced a mathematical theory of reflex conditioning (Grossberg and Schmajuk 1987; Grossberg et al. 1988), which rested upon a more complex variant of the gated dipole, called REcurrent Associative gated Dipole (READ). This new model extended the previous with functionality for memorizing emotions by employing all three elementary mechanisms (neuron activation, signal transmission, and synaptic memory). The READ network contains additional neurons and synapses endowing it with *long-term emotional memory*, needed to store associations between external stimuli and emotions they provoked in the past. In effect, this is the way to model behaviour in which an organism learns to avoid harmful environmental influences as they get associated with unpleasant emotions. In contrast, positive emotions connect with external events delivering some kind of utility. Some associations are easier to build than others are: for example, people learn more quickly to fear predatory animals than technical accidents—the former have been with us for millions of years, while the latter are recent history. READ's mathematics is general enough to deal with all of these connections.

READ stores past emotions in a long-term memory because it contains more neurons and synapses than the simpler dipole. Alongside Eqs. (5.5)–(5.14), the following new equations are added:

$$\frac{dy_7}{dt} = -y_7 + G[y_5]^+ + L\sum_{k=1}^{S} y_k z_{7k} \qquad (5.15)$$

$$\frac{dy_8}{dt} = -y_8 + G[y_6]^+ + L\sum_{k=1}^{S} y_k z_{8k} \tag{5.16}$$

$$\frac{dz_{7k}}{dt} = y_k(-h_1 z_{7k} + [y_5]^+) \tag{5.17}$$

$$\frac{dz_{8k}}{dt} = y_k(-h_1 z_{8k} + [y_6]^+). \tag{5.18}$$

Here G, L are real and positive constants. Equation (5.15) describes a neuron whose behaviour is driven by two factors. The first is the emotion $[y_5]^+$ from the leading channel, and the second is the sum $\sum_{k=1}^{S} y_k z_{7k}$, which represents the joint action of all conditioned stimuli. Here, S is their total number.

In experiments, these stimuli can be just like those used by Pavlov—bells, rings, lights, and food, given to hungry laboratory animals. In the life of humans, the stimuli are various goods and services with their specific attributes. In their role of consumers, people pay attention also to factors such as prices, price expectations, own financial constraints, sellers' and suppliers' reputations, public norms, etc. With experience, one learns how to deal with newly emerging products and services, and gradually develops habits for using them. Suppliers, in turn, study their customers carefully and try to seduce them with interesting offers and advertisements. John Watson, founder of behaviourism and author of the stick-and-carrot metaphor, has concluded in the distant 1922 that,

> The consumer is to the manufacturer, the department stores and the advertising agencies, what the green frog is to the physiologist (DiClemente and Hantula 2003).

We may dislike the idea of denigrating our social life by comparisons involving animals, but we should not overlook two major points in that statement. First, the human being is treated as an organism receiving stimuli and producing responses. Secondly, there is the opportunity to teach and cultivate the customer, who is able to develop new needs for newly emerging temptations. Equations (5.15)–(5.18) are good enough to explain these effects.

Back to the technical discussion, Eq. (5.16) is similar to Eq. (5.15). Equations (5.17) and (5.18) model long-term memories and coincide with Eq. (5.4) as discussed in Sect. 5.4.1. In particular, instead of y_k in Eq. (5.4) we have $[y_5]^+$ in Eq. (5.17), which is the emotion o_1 in the leading channel.

Finally, the neutral stimulus from outside must be linked with a visceral unconditioned stimulus to turn into a conditioned stimulus. That is how the Pavlovian dog learns to associate lighting a lamp with receiving some food. READ can account for this effect due to a modification of Eqs. (5.5) and (5.6), shown in Eqs. (5.19) and (5.20) below. Let us designate signal J (see Fig. 5.4) to be an unconditioned[2] stimulus (hunger, thirst, etc.). It must be submitted to the dipole jointly with a conditioned stimulus whose role is played by terms My_7 and My_8 in the following equations:

$$\frac{dy_1}{dt} = -y_1 + I + J + My_7 \tag{5.19}$$

$$\frac{dy_2}{dt} = -y_2 + I + My_8. \tag{5.20}$$

Here M is a real and positive constant. Now it becomes possible for this neural network to model consumer behaviour in general. Let a human organism feel certain need, which is represented by J in Eq. (5.19). That person acquires a product or service to satisfy it. During the act of consuming, the relevant dipole in the brain receives a signal y_k in $\sum_{k=1}^{S} y_k z_{7k}$ as per Eqs. (5.15) and (5.16). Neurons y_7 and y_8 get affected and propel the influence to neurons y_1 and y_2 as seen from Eqs. (5.19) and (5.20). Due to the feedback loop, all further neurons along the two channels perceive the change. Depending on the product's qualities and ability to meet the particular need, it provokes emotion of satisfaction or disappointment in one of the channels.

Memories z_{7k} and z_{8k} remember that emotion, as described by Eqs. (5.17) and (5.18). In short, while consuming a product, the client develops an opinion about its ability to satisfy a particular need, which process is a specific form of reflex conditioning. The READ neural model captures it entirely.

5.7 A Mix of Old and New Decision Science

By necessity, the discussion so far has been mostly theoretical. Now I adopt a pragmatic approach and show how the new scientific method can be related to some of the established concepts in decision science. To this end, I describe briefly a laboratory experiment in consumer behaviour and analyze it with the apparatus outlined in this chapter.

Quite unexpectedly, some of the important findings will lend themselves to interpretation involving Kahneman and Tversky's work. In particular, the method of computational neuroscience will elicit effects, which have previously been examined only in the framework of prospect theory.

---

[2]In other cases J may be treated as conditioned stimulus, and the mathematical formalism would not be changed. It is essential that various configurations of gated dipoles can model conditioned reflexes, associated with different needs—from more basic, like physiological needs, to more sophisticated, like cultural or aesthetic needs. Concrete circumstances would dictate if J should represent an unconditioned, primary conditioned, or secondary conditioned stimulus.

In his 2011 book, Kahneman paid some tribute to other theories that had less influence than prospect theory, but had made more accurate predictions in certain cases than it had. Its advantage, however, was its greater overall simplicity. In this section, our main research weapon is Grossberg and Gutowski's affective balance theory and its central element, the gated dipole. This theory is more complex and in addition, more general. It explains effects such as the paradoxical preference reversals, the gambler's fallacy, many framing effects, and risk aversion (Grossberg and Gutowski 1987), at least some of which are beyond the ability of prospect theory. Therefore, it is hardly surprising that a more general theory can arrive at the same conclusions as another theory.

Now I take a bold step further. An argument can be developed for gracefully pensioning out prospect theory and utility theory altogether, at least in decision science. It starts with the observation that theoretical economics benefits greatly from the concept of utility function, but putting the question "how people choose between options" in the centre of research calls for a deeper analysis. It becomes immediately apparent that a description of the agent by a utility function is highly problematic. In the 21st century, it is untenable to claim that the brain "computes the value/utility" of options and then simply picks the one with the highest number. Thus far in the book we already know that each decision is extremely sensitive to the prospects' formulations and context. Moreover, even this position is not sufficiently accurate because it turns out to be not radical enough. A growing number of studies, including a comprehensive 2011 review on all the important achievements in the field by Vlaev et al. (2011) showed that the scientific community is reaching a consensus around the following conclusion: *Choice depends above all on direct and* ad hoc *comparison between options, without any computation of value.*

In this situation, if there is room left for the utility function, it is shrinking. Modern decision science, and even economics, tends to use this concept mostly in a limited statistical sense because it is indeed pointless to analyze any particular decision by resorting to utility functions. Only analyses of aggregations of decisions, especially taken by samples of people rather than by individuals, may still benefit from this "old workhorse". In the next sections, we bid farewell to the utility function but do it in a respectful manner.

5.7.1 An Experiment in Economic Psychology

With my colleagues Henrik Egbert, Stefan Pulov, and Kalin Georgiev, we conducted an experiment in economic psychology (Mengov et al. 2008). As a detour, with my graduate student Svetla Nikolova we introduced a utility function derived from the differential equations of the three cognitive mechanisms described earlier (Mengov and Nikolova 2008). This was a kind of a methodological stroll—we sought to examine how a new research philosophy could go out of its way to produce a conceptually outdated instrument and see how useful it could be. To this end, we applied the gated dipole theory and calibrated the new utility function for each person separately. Then we analyzed the data over the entire sample of participants. We discovered an intriguing finding about customers' attitudes towards a particular service in a European country.

The experiment was computer-based and the participants assumed the role of consumers of a fictitious service, bearing resemblance to mobile phone operators. In each of 17 rounds, a participant had to choose between retaining the current supplier or changing it with a competitor, based on information about expected and actual prices to be paid. An essential element of the study was measuring the consumer satisfaction, self-assessed and reported immediately after each round. On a computer screen the subjects had to express their feelings about how they were treated in the current round by their supplier in this Lickert-type scale: "Extremely disappointed (-4)—Very disappointed (-3)—Disappointed (-2)—More disappointed than satisfied (-1)—As much satisfied as disappointed (0)—More satisfied than disappointed (+1)—...—Extremely satisfied (+4)". These adverb-adjective compounds were chosen such as to make the measurement scale interval by ensuring that the segments between neighbouring points were equidistant in line with Cliff's (1959) multiplicative rule. Finally, the timing of each mouse-click was recorded alongside the action associated with it, and was later used in calibrating the system of differential equations.

In particular, a customer had to choose between one of two suppliers A and B, based on the expectation that at the end of the period/game-round they would likely pay an amount P_a (price advertised). The actual price to be paid, however, was often different. We denote it P_f (final price). A difference $\Delta P = P_a - P_f$ occurred. In most cases, a saved amount ΔP^+ would provoke satisfaction; similarly, an extra charge ΔP^- would be disappointing. However, it is quite natural that a discount felt to be too small might also provoke disappointment, and an unexpectedly small extra charge might cause a mild satisfaction. The gated dipole model is capable of accounting not only for choice behaviour in general, but also for such delicate effects in particular. Its equations are the same as already discussed, with the exception of (5.19), (5.20), which were adapted to become:

$$\frac{dy_1}{dt} = -y_1 + P_a + \delta \Delta P^+ + M y_7 \tag{5.21}$$

$$\frac{dy_2}{dt} = -y_2 + P_a + \delta \Delta P^- + M y_8. \tag{5.22}$$

In these equations δ is a parameter, scaling the relative influence of the price difference, and is calibrated alongside the other parameters. A variety of prices P_a, P_f, was shown to 129 participants, and their reactions were recorded. We mapped all that data onto the gated dipole by selecting suitable values for constants δ, M, b_1, b_2, c_1, c_2, G, L, h_1, and some others for each person.

Because the system of differential equations cannot be solved analytically, we implemented stochastic optimization to find an approximate solution. The particular optimizing method used was a fast version of Simulated Annealing. It generated random values within certain intervals for the constants and produced many solutions at each iteration. An error function containing all differences between the empirical data (participants' satisfactions and choices), and the relevant system variables was computed for each solution. The currently best solution was retained. The intervals for all constants shrank slightly according to a rule, favouring the best solution so far. In theory, continuing this process for infinitely many iterations with infinitely small interval shrinkages would lead to the error function's global minimum. In practice, we did about 30,000 solutions for each participant to achieve 85–95 % correct predictions on the entire sample of 129 people (Mengov et al. 2008). The differential equations were numerically solved by a Runge–Kutta–Fehlberg 4–5 method.

Because we expected asymmetric human reactions to gains and losses, in Eqs. (5.7)–(5.8) we introduced two sets of constants b_1 and b_2, and also c_1 and c_2 instead of simply b and c. Had people reacted differently to gains, associated in this model with constants b_1 and c_1, than to losses, associated with b_2 and c_2, then one could expect statistically significant differences of the type $b_1 \neq b_2$, $c_1 \neq c_2$, or even both.

And indeed, for the sample of our participants we found a highly significant difference of the kind $b_1 > b_2$, shown by both the Wilcoxon Matched Pairs Test and the Sign Test (Mengov and Nikolova 2008).The other hypothesis: $c_1 = c_2$, was not rejected, meaning that the dipole channels were found symmetric enough with regard to constant c. Taken in conjunction, these two results are remarkable in more than one respect. The first has to do with a computational issue, the second concerns human behaviour, and finally there are some more general considerations. Let me discuss them one by one starting with the technical matter.

A number of studies (Marder and Taylor 2011; Prinz et al. 2004) have shown that both neural network models and single neuron models with only a handful of parameters—and in this study we used 13, which is a lot—exhibit a particular kind of instability: indistinguishable model behaviours can arise from multiple sets of parameter values. In fact, any neural network researcher is quite familiar with this phenomenon because it is at the heart of network training. Two networks with identical structures and equally good performance usually have reached two quite different values for each parameter (called connection weight). In mathematical biology, the same effect is less welcome and is known as the "degeneracy problem", meaning that one and the same biological outcome may be reached via different evolutionary paths.

A stochastic optimization method such as simulated annealing, in which every new solution is based on random generation of uniformly distributed parameter values, is a generator of parameter instability, or "degeneracy", par excellence. Therefore, a failure to reject hypotheses such as $b_1 = b_2$ and $c_1 = c_2$ when these constants are optimized in conjunction with many more like them would be naturally expected. Under these adverse circumstances, finding a statistically significant difference of the type $b_1 > b_2$ is all the more amazing. It proves that there is indeed a robust asymmetry in the functioning of the dipole's two channels. The nature of this asymmetry was another surprise.

5.7.2 An Unorthodox Utility Function

Let me now explain what the established parameter asymmetry means for the observed human behaviour. Omitting the technical discussion in (Mengov and Nikolova 2008), here I take its conclusion, which is that statistically, i.e. as an average over the entire sample of 129 people, the following inequality holds:

$$o_1(x) > o_2(x). \tag{5.23}$$

We may recall that o_1 and o_2 are functions of neural activities, directly related to the actual customer satisfaction and disappointment. Inequality (5.23), alongside Eqs. (5.13)–(5.14), states that in the experiment, i.e. for the particular sample of people, *satisfaction with the suppliers has been greater than disappointment—on average*. This finding is bizarre because the experimental conditions did not appear to have favoured such an effect. They were designed to be balanced, in the sense that the number of people who received discounts was equal to the number of people who endured extra charges, and the total sums of quantities ΔP^+ and ΔP^- were also equal. Thus, all gains and losses canceled each other statistically. Why should then people be—on average—more satisfied than disappointed?

The paradox can be fully understood by putting it in the perspective of prospect theory. Kahneman and Tversky famously established that people's utility curve was generally steeper for losses than for gains. In particular, the average intensity of the positive emotion from discount amounting to ΔP should be smaller than the average intensity of the negative emotion from extra charge ΔP, and *not* vice versa. Therefore, our result was tantamount to anomaly in the shape of the utility function, at least at first sight.

To understand what had happened we had the following idea—why not derive analytically a new utility function from the gated dipole equations and analyze it? If we had a formula, we could fully specify it for each person from the sample because numerical values were available for all the constants due to simulated annealing. Within the frame of a thought experiment, we introduced some simplifying assumptions into Eqs. (5.5)–(5.22) allowing us to solve them jointly at equilibrium (Mengov and Nikolova 2008). The logic of that was to capture the statistical effects accumulated in the constants without bothering about the system dynamics. Such a thing cannot be achieved by examining a living organism, but that is what thought experiments and models are all about.

Under the stated assumptions and without taking the exercise too seriously, we set the derivatives to zero, and with some assistance from the software package Mathematica solved the resulting algebraic equations. The utility function derived in this way was:

$$V(\Delta P) = \begin{cases} y_5(\Delta P) - y_5(0), & \text{if } \Delta P \geq 0 \\ -y_6(-\Delta P) + y_6(0), & \text{if } \Delta P < 0, \end{cases} \tag{5.24}$$

where

$$y_5(\Delta P) = \frac{b_1 c_2 P_a \hat{P} - b_2(-b_1 \delta \Delta P + c_1 P_a \hat{P})}{b_1 b_2 + c_1 c_2 P_a \hat{P} + (b_2 c_1 + b_1 c_2) P_a \hat{P} + b_1 b_2 (P_a + \hat{P}) + (b_1 c_2 P_a + b_2 c_1 \hat{P})},$$

$$y_6(\Delta P) = \frac{-b_1 c_2 P_a \hat{P} + b_2(b_1 \delta \Delta P + c_1 P_a \hat{P})}{b_1 b_2 + c_1 c_2 P_a \hat{P} + (b_2 c_1 + b_1 c_2) P_a \hat{P} + b_1 b_2 (P_a + \hat{P}) + (b_1 c_2 P_a + b_2 c_1 \hat{P})},$$

and for compactness, the substitution $\hat{P} = P_a + \delta \Delta P$ was introduced. The expressions for $y_5(\Delta P)$ and $y_6(\Delta P)$ shown here are simpler than those in (Mengov and Nikolova 2008) because due to experience, we set some constants equal to one.

The utility function (5.24) comprises a fictitious, statistical approximation of each person's overall experience with utilities in general, and with the gains and losses in our laboratory study in particular. To differentiate it from its established counterparts in utility theory and prospect theory, we denote it with V rather than with U. The adjustments $y_5(0)$ and $y_6(0)$ were introduced to offset nonzero quantities, and naturally established a zero reference point.

The *rhs* expressions for $y_5(\Delta P)$ and $y_6(\Delta P)$ in Eq. (5.24) are not amenable to intuitive interpretation, which makes them look more typical for some theoretical field in the natural sciences than for a social science. All the same, the entire function assumed a form analogous to the utility function in cumulative prospect theory, as shown in Chap. 3, Eq. (3.12). A comparison between the two is shown in Figs. 5.5 and 5.6. The left plot in the former figure presents the classical qualitative characterization of the function with its steeper branch for losses. The right plot in Fig. 5.5 shows the function as per Eq. (3.12), computed with constants α, β, and λ from Tversky and Kahneman's 1992 experiment.

Next, Fig. 5.6 plots typical functions for four participants in our experiment computed with Eq. (5.24). In shape, they more or less agree with prospect theory. The top-right plot in Fig. 5.6 and the curve from cumulative prospect theory (Fig. 5.5, right) look—up to a scaling coefficient—so much alike that one forgets, and can hardly believe how different their methodological foundations are.

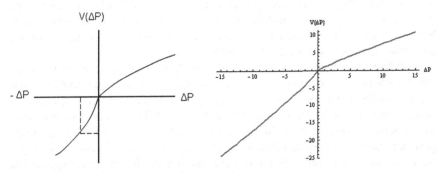

Fig. 5.5 Utility functions in prospect theory (*left*) and cumulative prospect theory (*right*). Here, the gains and losses are ΔP [Reprinted from Mengov and Nikolova (2008), with the permission of The Bulgarian Academy of Sciences Publishing House.]

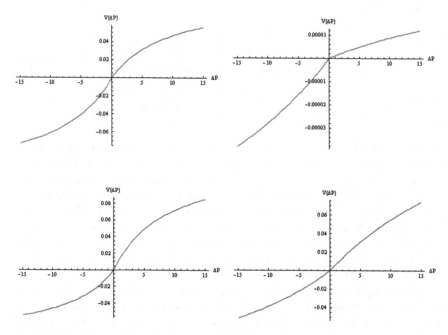

Fig. 5.6 Gated dipole utility functions by Eq. (5.24) of four participants in the experiment [Reprinted from Mengov and Nikolova (2008), with the permission of The Bulgarian Academy of Sciences Publishing House.]

The top-left plot shows another example, appearing again very much in support of prospect theory. However, those two were not typical participants.

The majority of people reacted according to Eq. (5.23), i.e. for them satisfaction exceeded disappointment, which meant that their utility curves by Eq. (5.24) looked like the two bottom plots in Fig. 5.6: *They were steeper for gains, not for losses.* On average, receiving ΔP was somehow more influential than losing ΔP. Were those people a minority, this finding would not have been problematic. However, they were the majority, and re-checking all calculations of all utility functions and statistical tests times and again failed to reveal an error. Then my coauthor Svetla Nikolova suggested an explanation, which was later confirmed in many post hoc interviews and discussions with the subjects.

It all had to do with the wider context of the study. In 2007, in the European country where we conducted the experiment, there were only two mobile phone providers with digital networks. The second operator had appeared only a couple of years earlier. For quite some time before that, a single provider had enjoyed a monopolistic market position. No one ever proved its abuse, but most customers genuinely held the belief that they were often mistreated. Extra charges were perceived as the norm while discounts were unheard of. In this environment, our experimental design offered a fair treatment to our subjects and that was unusual to them. They were accustomed to a different reality, which prospect theory would describe as shifting the reference point towards losses. Exactly that is indicated with dashed lines in Fig. 5.5, left plot. All of a sudden, a systematically occurring steeper curve for gains was not only possible, but even natural.

That was a curious research experience. The primary goal of the study was to relate an established model from mathematical neuroscience with actual consumer behaviour. In that, we succeeded as demonstrated in detail in (Mengov et al. 2008). Indeed, our model managed to predict people's choices between providers A and B for the next round with 95 % accuracy on the calibration sample and 87 % on the test sample. This comprised a modest improvement on the classical econometric techniques.

As a detour, we derived a probably fanciful formula for utility. All the same, it could be used on a person-by-person basis and its constants had clear neurobiological meaning—two qualities, which at least partially offset its somewhat excessive complexity. Finding connections with prospect theory,—the dominant yet phenomenological theory in behavioural economics—helped us discover an unexpected bias in the mind of the typical consumer of cell phone services. This was an intriguing small discovery and an added bonus.

Yet, I stress again that the neurobiological approach outlined in this chapter has a potential that cannot be reduced to the narrow concepts of old theories. If this is done sometimes, it is only to establish connections with previous work and show some continuity. The gist of the new method is developing real-time neural network models, explaining the dynamics of cognitive-emotional interactions that are central to decision-making. All of this is beyond the state-of-the-art decision science.

5.8 We Are Homo Aequanimus

Finally, it is time for a few general thoughts about utility. Discussing it, I find it worthwhile to compare the computational neuroscience paradigm with the mainstream scientific approaches. Over the last decades, a multitude of studies in economic behaviour has involved psychological methods; yet these have generally been rooted in the assumption that the sole purpose of the agent is to maximize subjective utility. Scholars have almost routinely discovered, however, that under a variety of circumstances people persistently do not behave in this way even in cases with very simple tasks. The many kinds of paradoxical preference reversals are the epitome for this line of research. Ad hoc theories, among them some quite ingenious, have devised numerous commonsensical explanations. Taken together, they appear disparate, though.

Already in the fMRI era, we have seen studies claiming to have discovered the particular neurons that compute value or utility, and have sought to establish whether it was relative or absolute value/utility (Padoa–Schioppa and Assad 2006, 2008). In the view maintained in this book, such investigations have taken a path a bit too easy. In the already distant 1987, Grossberg and Gutowski's affective balance theory (Grossberg and Gutowski 1987) introduced a much more plausible framework for reasoning:

> This analysis [...] provides an explanation of why individuals often do not act to maximize subjective value, even in simple situation where cognitive complexity is minimal. [This is because] the mechanisms that prevent maximization have manifest adaptive functions. These mechanisms are not designed to maximize subjective value. Rather, they are designed to control the emotional processes that regulate reinforcement, incentive motivation, and affectively modulated attention shifts.

Thus, our mind does not engineer utility maximization. Making our choices, we seek emotional balance. We are not Homo Economicus. We are Homo Aequanimus—someone who resorts to the opponent processing of their emotions to guide them through the flow of daily events. It is advantageous to pursue equanimity. Once a negative affect has made us reassess a situation, it has to abate quickly, and that is achieved by the neurotransmitter action in the opposite dipole channel. The lonely opera fan from the beginning of this chapter cannot wait forever, but must decide to do something else. The problem of the gambler who would not stop betting is that his channel for negative emotions has become depleted, and is no longer effective in preventing the disaster.

Another time, a pleasant feeling follows a success—how we got there, we had better remember, but euphoria should not last too long either. In short, opposite emotions alternate, driven by events in the environment, and steer our course.

That is why decisions are not governed by utility maximization, but by a cognitive-emotional mechanism with a somewhat different purpose. At its heart is the goal of maintaining or quickly restoring the emotional balance—the very basis for adaptive and advantageous behaviour since millions of years. Of course, satisfaction usually comes after we have maximized some kind of utility. With humans, this has been the product of substantial cognitive effort. The more plotting and strategizing, the greater the returns—ceteris paribus and with some luck. When, however, that utility gets too much in the way of our wish for inner peace, the latter eventually prevails, and the aspirations of Homo Economicus within us must subside.

References

Birnbaum, M. H. (1999). Paradoxes of Allais, stochastic dominance, and decision weights. In J. Shanteau, B. A. Mellers, & D. A. Schum (Eds.), *Decision science and technology: Reflections on the contributions of Ward Edwards* (pp. 27–52). Norwell: Kluwer Academic Publishers.

Birnbaum, M. H. (2008). New paradoxes of risky decision making. *Psychological Review, 115,* 463–501.

Busemeyer, J. R., & Townsend, J. T. (1993). Decision field theory: A dynamic-cognitive approach to decision making in an uncertain environment. *Psychological Review, 100*(3), 432–459.

Cliff, N. (1959). Adverbs as multipliers. *Psychological Review, 66,* 27–44.

DiClemente, D., & Hantula, D. (2003). Applied behavioural economics and consumer choice. *Journal of Economic Psychology, 24,* 589–602.

Grossberg, S. (1967). Nonlinear difference-differential equations in prediction and learning theory. *Proceedings of the National Academy of Sciences of the United States of America, 58*, 1329–1334.

Grossberg, S. (1969a). On the global limits and oscillations of a system of nonlinear differential equations describing a flow on a probabilistic network. *Journal of Differential Equations, 5*, 291.

Grossberg, S. (1969b). Embedding fields: A theory of learning with physiological implications. *Journal of Mathematical Psychology, 6*(2), 209–239.

Grossberg, S. (1972). A neural theory of punishment and avoidance, II: Quantitative theory. *Mathematical Biosciences, 15*, 253–285.

Grossberg, S. (1998). Birth of a learning law. *INNS/ENNS/JNNS Newsletter, 21*, 1–4.

Grossberg, S. (2009). Cortical and subcortical predictive dynamics and learning during perception, cognition, emotion, and action. *Philosophical Transactions of the Royal Society of London,* (special issue "Predictions in the brain: Using our past to generate a future") *364*, 1223–1234.

Grossberg, S. (2013). Adaptive resonance theory: how a brain learns to consciously attend, learn, and recognize a changing world. *Neural Networks, 37*, 1–47.

Grossberg, S., & Gutowski, W. (1987). Neural dynamics of decision making under risk: Affective balance and cognitive-emotional interactions. *Psychological Review, 94*(3), 300–318.

Grossberg, S., Levine, D., & Schmajuk, N. (1988). Predictive regulation of associative learning in a neural network by reinforcement and attentive feedback. *International Journal of Neurology, 21–22*, 83–104.

Grossberg, S., & Pilly, P. (2008). Temporal dynamics of decision-making during motion perception in the visual cortex. *Vision Research, 48*, 1345–1373.

Grossberg, S., & Schmajuk, N. (1987). Neural dynamics of attentionally-modulated Pavlovian conditioning: Conditioned reinforcement, inhibition, and opponent processing. *Psychobiology, 15*(3), 195–240.

Hodgkin, A. L., & Huxley, A. F. (1952). A quantitative description of membrane current and its applications to conduction and excitation in nerve. *Journal of Physiology, 117*, 500.

Kuhn, T. (1962, 1970). *The structure of scientific revolutions*. Chicago: The University of Chicago Press.

Lettvin, J., Maturana, H., McCulloch, W. S., & Pitts, W. (1959). What the frog's eye tells the frog's brain. *Proceedings of the Institute of Radio Engineers, 47*, 1940–1951.

Litt, A., Eliasmith, C., & Thagard, P. (2008). Neural affective decision theory: Choices, brains, and emotions. *Cognitive Systems Research, 9*, 252–273.

Marder, E., & Taylor, A. L. (2011). Multiple models to capture the variability in biological neurons and networks. *Nature Neuroscience, 14*(2), 133–138.

McCulloch, W. S., & Pitts, W. (1943). A logical calculus of the ideas immanent in nervous activity. *Bulletin of Mathematical Biophysics, 5*(4), 115–133.

McCulloch, W. S., & Pitts, W. H. (1947). How we know universals. *Bulletin of Mathematical Biophysics, 9*, 127–147.

Mengov, G., Egbert, H., Pulov, S., & Georgiev, K. (2008). Affective balances in experimental consumer choices. *Neural Networks, 21*(9), 1213–1219.

Mengov, G., & Hristova, E. (2004). The human factor in decisions under risk in industrial systems. *Automation and Informatics, 38*(2), 11–13.

Mengov, G., & Nikolova, S. (2008). Utility function derived from affective balance theory. *Proceedings of the Bulgarian Academy of Sciences (Comptes Rendus), 61*(12), 1605–1612.

Padoa–Schioppa, C., & Assad, J. A. (2006). Neurons in orbitofrontal cortex encode economic value. *Nature, 441*, 223–226.

Padoa–Schioppa, C., & Assad, J. A. (2008). The representation of economic value in the orbitofrontal cortex is invariant for changes of menu. *Nature Neuroscience, 11*, 95–102.

Prinz, A., Bucher, D., & Marder, E. (2004). Similar network activity from disparate circuit parameters. *Nature Neuroscience, 7*, 1345–1352.

Vlaev, I., Chater, N., Stewart, N., & Brown, G. (2011). Does the brain calculate value? *Trends in Cognitive Sciences, 15*(11), 546–554.

von Neumann, J., & Morgenstern, O. (1944, 1947, 1953). *Theory of games and economic behaviour*. Princeton: Princeton University Press.

Chapter 6
Choice by Intuition

6.1 Experimental Advances

As we saw in the preceding chapter, mathematical and computational neuroscience can successfully bridge the gap between decision science and the psychology of cognitive-emotional interactions. However, the limits within which neural models can explain empirical data from real economic choices remain virtually unknown. In this chapter, I present a sophisticated experiment, at once psychological, economic,[1] and computational. It deals with arguably the most interesting aspect of decision-making—the intuition. An additional point is that intuition is present in laboratory studies and in real markets in equal measure.

I will say in advance that the experiment outlined here ventured boldly into this uncharted territory and achieved some fascinating results. In particular, a model built around the READ network was shown to be able to forecast individual decisions in a complicated economic game on a person-by-person basis. The model performed extremely well with people who were guided by gut feelings; it failed though, with those participants who applied strategic reasoning. Because the adopted theoretical construction consisted almost entirely of equations describing neurons and their interactions, READ's success means that it can be regarded as a plausible model of the neurobiological substrate of primitive-intuitive thinking.

6.2 Rendezvous Between Theory and Experiment

A number of scientists have followed in the footsteps of Grossberg by adopting the method of combining and recombining the three differential equations to describe cognitive phenomena of increasing complexity. Usually, the outcome has been a

---

[1]The experiment is remotely related to both economic psychology and behavioural economics, but strictly speaking, belongs to neither of them.

© Springer-Verlag Berlin Heidelberg 2015
G. Mengov, *Decision Science: A Human-Oriented Perspective*,
Intelligent Systems Reference Library 89, DOI 10.1007/978-3-662-47122-7_6

complicated neural network consisting of a mosaic of gated dipoles and other building blocks. In particular, Daniel Levine proposed a number of models of that kind. Among them was one that successfully dealt with the "kiss-and-money finding"—the observation that decision makers distort substantially the small probability for gains when the trophy is emotionally charged (Levine 2012). The funny name comes from an experimental condition in which people's subjective probabilities were assessed in connection with the prospect of receiving money vs. the possibility to be kissed by their favourite movie star. The overestimation of the odds in the latter case was much stronger (Rottenstreich and Hsee 2001).

Other neural models by Levine involved Maslow's hierarchy of needs (Levine 2009), and the dual motives of selfishness and empathy in economic behaviour (Levine 2006). Earlier Leven and Levine (1996) introduced a somewhat tessellated neural network, used as the backbone for a theory about consumer motivation. While such constructions seem ad hoc-built, they are indeed thought provoking, especially when their authors manage to link them with fMRI findings involving the real brain anatomy.

All of these modelling efforts, however, have aimed at producing theoretical knowledge and have generally not been intended for direct applications. Trying to use one of Grossberg's or his disciples' theories to guide a novel laboratory experiment or R&D work usually leads to obstacles that call for introducing simplifications and adaptations, often sacrificing theoretical rigour. This is unavoidable for at least two reasons: first, the target domain—such as economic decision-making—may be quite different from the theory's territory of origin, in our case mathematical neuroscience. Secondly, in applied science, one often seeks to develop models for prediction that must deal with noisy data, contaminating factors, and artefacts.

In this way, however, the closing of a certain methodological loop is achieved. It looks like this: When empirical studies of a phenomenon attract attention, they may instigate the inception of a theory. That theory undergoes a validity check against the original data, and once sufficiently developed, can predict new phenomena. Sometimes the predicted findings may come out of studies, ignorant of the particular theory. A link between the two gets established only post hoc which is sad, because the predictions will have been sidestepped. In other cases, however, the theory serves as guiding light for new experimental work. Exactly that is what follows in this chapter.

6.3 The Omnium Bonum Economy

The experiment outlined here sought to establish to what extent people act in the affective-intuitive way, accounted for by the READ neural model, when they are put in a moderately complex economic situation. I describe it here following (Mengov 2014). In a number of rounds, subjects had to choose repeatedly one among four suppliers of a fictitious commodity called *omnium bonum* (a good for everyone, in Latin) seeking to obtain as much as possible of it. It had to be called by such a strange name to avoid direct associations with any real product or service. This is the standard approach in experimental economics as it eliminates potential confound effects (Zizzo 2013).

At the end of the game the accumulated units of the commodity were exchanged for real money. The participants had to orient themselves in an environment with insufficient information by subjectively developing in their minds adequate profiles of the suppliers and using them as choice determinant. The instruction was written so as to avoid any role-assigning in any concrete economic or business circumstances, again escaping possible confounds.

The four suppliers provided omnium bonum of equal quality, but were not always reliable in delivering it—they could offer certain amount, but often deliver less, or even more. Their profiles are summarized in Fig. 6.1, which shows how the supplier offering the most on average was also the least dependable. In essence, this was an implementation of the economic idea that higher profit goes hand in hand with more risk-taking. The subjects were unaware of this feature and could only discover it by trial and error.

The exact figures of the design were chosen such as to meet a number of requirements. First, there had to be real competition among the suppliers so that the participants could be facing real choices. Only that could make the experimenter's prediction efforts meaningful. Second, each supplier had to remain competitive throughout the game. Driving them out of business—for any reason—was undesirable as it would diminish the number of options. Conversely, no supplier ought to come close to monopolizing the market. Third, each supplier had to be in a position to form a distinct image in the eyes of the participant. One way of achieving this was by introducing the two-dimensional design shown in Fig. 6.1. Finally, the game had to be long enough for the suppliers to become recognizable, yet it could not be

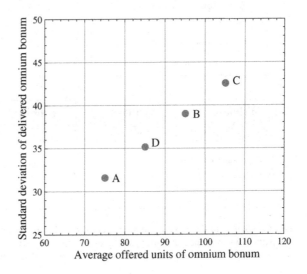

Fig. 6.1 Four suppliers offered and delivered different units of omnium bonum, whereby the riskiest (the one with the largest standard deviation) was most rewarding [Reprinted from Mengov (2013), with the permission of The Bulgarian Academy of Sciences Publishing House]

too long lest the participants got used to it to the point of routine or boredom, which would be two additional confounding effects.

Together with my graduate student Nickolay Georgiev, we introduced a number of design features ensuring that the above demands were met. In detail those are described in (Mengov 2014; Mengov and Georgiev 2013), while here only a general idea is given. The number of rounds was fixed at twenty. While suppliers often delivered omnium bonum either less or more than offered, these deviations summed up to zero for each sequence of five rounds, thus introducing a measure of implicit "fairness", should a supplier be chosen repeatedly. Of course no subject was obliged to do so, and moreover, was never given a hint about this arrangement. The order of delivery deviations was also unpredictable for the participant. No transaction costs were involved in abandoning one supplier for another. The four of them were put on the computer screen as shown in Fig. 6.2.

Further, in any given round at most two suppliers behaved similarly, e.g. by delivering exactly as having offered. In half of the experimental treatments a

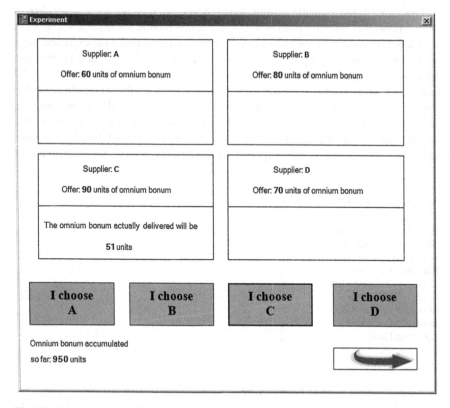

Fig. 6.2 Screen of the supplier offers and deliveries. In the example, sometime through the game the participant examined the offers and chose Supplier C with a mouse click. Immediately, the actual quantity of omnium bonum was 'delivered' and added to the total

continuous economic growth was simulated, implemented by raising the offered and delivered quantities by a number of units after each five rounds. Thus the subjects' motivation and involvement was maintained until the end of the game. Another design feature ensured that the supplier with the most modest offer was able sometimes to deliver more than what was offered by the frontrunner. Thus, a system of delicate balances created genuine competition as well as enough attractiveness for each supplier.

The other half of the treatments simulated economic growth followed by economic crisis. This was achieved by reducing the quantities of offered and delivered omnium bonum in the last five rounds by all suppliers. Again, all figures were chosen so as to maintain the above balances.

An essential element of the experiment was measuring the consumer satisfaction, self-assessed and reported immediately after each omnium bonum delivery. On a separate screen shown in Fig. 6.3 the subjects had to express their feelings about how they were treated in the current round. The implemented scale was identical with the one discussed in Chap. 5, Sect. 5.7.1. Finally, the timing of each mouse-click was recorded alongside the action associated with it, and was later used in calibrating the system of differential equations.

Immediately after the game, the software application demanded answers to a sequence of open and closed questions about the subject's strategy, tactical reasoning, and other issues, potentially relevant for the individual's decision-making style. Obviously these components were unrelated to the neural network model, but were included to investigate what factors could be linked with its potential predictive success.

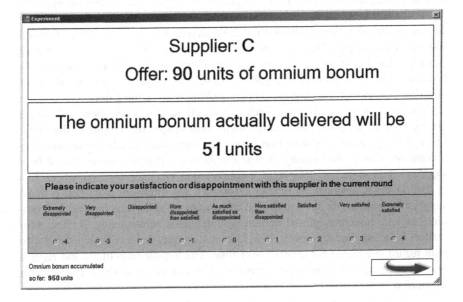

Fig. 6.3   Screen for self-assessed emotion after a delivery of omnium bonum

6.4 A Model for Each Individual

Having data about a person's reactions in 20 rounds of a game is not enough for statistical analysis, yet it may be quite informative when one has—or one believes that one has—a solid theory to explain it. READ, with its little more than a dozen neurons, turned out to be a sufficient theoretical foundation for understanding intuitive behaviour at the individual level.

Let me continue this discussion with an illustration. Figure 6.4 shows the specific realization of the neural network for choosing one among four options. While the subject watches the screen with the offers, neurons y_A, y_B, y_C, and y_D become active one by one. They interact with the dipole neurons to assimilate what is happening.

Imagine that you are the player and are just starting the game. In the very first round, all suppliers present their offers, each looking like a reason for delight. You examine every corner of the screen and gradually begin to feel eager to grab at the most generous-looking Supplier C. Then you probably recall that promises would be kept liberally, but at that point this means nothing to you. After a couple of glances at each option, a decision is taken to go for C. All of this, and how it continued for another person who really played the game, is illustrated in Fig. 6.4, the plots for neuron activities and emotional memories.

The jittery signals y_1 and $[y_5]^+$ at the very beginning (see the first 70 s. plots) reflect an eyeballing process that promised potential gains and thus caused satisfaction. A closer look at z_{7C}, the memory for positive emotion, shows how it jumped a little at the onset of the game, while z_{8C} did not change. Then Supplier C delivered much less than promised, satisfaction $o_1 = [y_5]^+$ fell to zero, the opposite dipole channel took advantage by activating its neurons, with memory z_{8C} storing the disappointment. Shortly before the 20th second, however, the round finished and a new one began. This time the satisfaction from eyeballing was less intense and hesitation lasted longer. Again, Supplier C was given a chance, only to dispense a second disappointment. All the same, our player turned out to be an obstinate person, not budging but quickly picking C for a third time.

With this experimental design, each participant generated a sequence of unique history with all suppliers, thus letting them form unique reputations in the mind. The following operational definition for *positive (negative) supplier reputation* was introduced: This is *the memory of past satisfaction (disappointment) caused by a supplier*. It is stored in long-term emotional memory) z_{7k} (or z_{8k}) where $k = A, B, C, D$ and is described by Eqs. (5.17)–(5.18). The important point here is that a supplier's reputation is not determined merely by the difference between the omnium bonum quantities offered initially and delivered subsequently. Rather, what matters is the emotional memory, individual and unique, that a client has developed with regard to a particular supplier over time. The logistic ratio of z_{7k} and z_{8k} is a way of accounting for the total reputation, not just the positive or the negative memories.

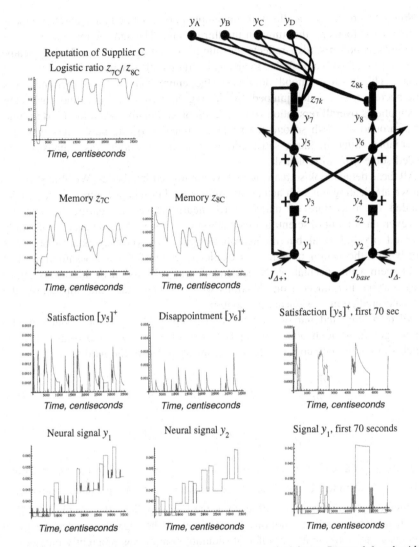

Fig. 6.4 Dynamics of the neural model for an anonymous participant. *Bottom left* and *middle plots* signals y_1 and y_2 are the response to the incoming stream of offers, deliveries, and to the participant's eyeballing before choosing. In the example, the person kept choosing Supplier C in the first three rounds and received extra omnium bonum in the third, which is reflected in the two "first 70 s" plots (*third column, below the neural circuit*). There, y_1 produced "ripples" at the onset of each round and then jumped around the 4600th centisecond (46th second) due to the surplus delivered. Around the 2000th centisecond, the corresponding $[y_5]^+$ signal shows that eyeballing four positive options can cause satisfaction, almost as intense as that of the actual lavish treatment. *Upper-left plots* the memory for positive emotions z_{7C} initially rose negligibly due to eyeballing, and then increased around the 50th second after the generous delivery in the third round. In contrast, the memory for negative emotion z_{8C} rose steeply in the first round and remained high in the next due to disappointingly unfulfilled promises. The supplier's dynamic reputation was defined by the logistic ratio z_{7C}/z_{8C} reflecting the two memories' joint action. [Reprinted from Mengov (2014), with the permission of Elsevier.]

Now let us return to the illustrative example in Fig. 6.4 and have another look at the first 70 s, focusing this time on the third round. The satisfaction curve—from the 43rd second onwards—reflects the jittery eyeballing process, which is seamlessly continued with a peak due to an apparently plentiful delivery around the 46th second. This positive mood dominated the entire round. The experience was remembered in z_{7C}, which jumped up, shifting the z_{7C}/z_{8C} balance and improving the supplier's overall reputation (Fig. 6.4, top plot). Finally, when the fourth round began around the 56th second the usual examination of the new offers was less pleasurable—after so much satisfaction the neurotransmitter was exhausted and needed some time to replenish.

All these details look so realistic that we are tugged into belief. We almost forget that we are dealing not with actual measurements, but with a computational model founded on a mixture of mathematical neuroscience and daring speculation. However, an amount of scientific imagination was indeed needed because this study involved psychological but no neurobiological data, yet they had to be made tractable by the neural model. This challenge called for some assumptions that might seem a bit speculative. Nonetheless, results discussed in the next sections show that the introduced simplifications were justified—they did not hamper the adequate prediction of subjects' behaviour.

All data were fed into READ via two channels. Information about omnium bonum got in through neurons y_1 and y_2, described by the following equations, which show how the supplier offers were incorporated in the model:

$$\frac{dy_1}{dt} = -y_1 + J_{base} + \delta J_{\Delta+} + \delta_1 J_{eye}^{(k)} + My_7 \tag{6.1}$$

$$\frac{dy_2}{dt} = -y_2 + J_{base} + \delta J_{\Delta-} + My_8. \tag{6.2}$$

Term J_{base} is the baseline signal, also known as tonic signal in the gated dipole. Term $J_{eye}^{(k)}$ with index $k = A, B, C, D$ accounts for the omnium bonum commodity q_k, offered by and associated with each of the four suppliers. By submitting $J_{eye}^{(k)}$ to the channel for positive emotions I implemented the idea that under the game circumstances, looking at any offer of omnium bonum was inherently pleasant.

Signal $J_{eye}^{(k)}$ is greater than zero only while the subject eyeballs the offers on the screen and deliberates which one to choose. All quantities of the commodity are scaled to become $J_{eye}^{(k)} = q_k/10\max(q_k)$ and not exceed 10 % of the maximum neural signal value of 1. The two signals are related:

$$J_{base} = \min\left\{J_{eye}^{(k)}\right\}_k. \tag{6.3}$$

Equations (6.1)–(6.3) embody a number of assumptions that had to be introduced in order to link neural and economic variables. The logic of Eq. (6.3) is that J_{base} is neural activation caused by the weakest offer. This may not be the only way to establish the baseline signal but it rests upon sufficient common sense: Examining the four offers, a participant is likely to view the weakest as a reference point for assessing the potential benefit from the remaining three. An alternative approach would have used some form of averaging over the four, but that would be an unnecessary complication. Terms $J_{\Delta+}$ and $J_{\Delta-}$ reflect a positive difference (more omnium bonum delivered than promised) and a negative difference (less delivered than promised), and are also scaled down to $\Delta q_k / 10 \max(\Delta q_k)$. Equations (6.1)–(6.2) contain positive constants δ, δ_1 which due to previous experience are fixed in advance, and M, which is determined together with some other constants in a stochastic optimization procedure.

Implementing the eyeballing signal $J_{eye}^{(k)}$ in the computational model involved another amount of scientific speculation. It was postulated that in each eyeballing-and-deliberation period all four options were sufficiently well understood. In reality, it may have taken several glances at each of them, and maybe more attention was paid to one than to another, especially after some initial experience. The model assumed that in each round the subject divided her/his attention (i.e. time) among the four suppliers *equally*, and cast exactly *three* glances at each of them *at random*.

Thus, the eyeballing period was split in 12 equal parts, during which $J_{eye}^{(k)}$ propelled the offers inwards as per Eq. (6.1). This happened in synchrony with signals $y_A, \ldots, y_D$, switched on and off accordingly. Of course it cannot be claimed that such was the real sequence of events, but only that this was a reasonable way to model how the options were adequately grasped.

The adopted twelve-look scheme was an efficient way to avoid modelling the visual perception process, which would have been very interesting and at least as complicated. Now the scheme simply implied that the time a participant needed for examining offers and taking a decision had to be divided by 12. This was a plausible assumption with regard to the overwhelming majority of response times, but needed a closer look in the boundary case of the quickest responses. In particular, the shortest recorded interval was 530 ms, which happened twice and was achieved by one and the same subject towards the end of the game. This number divided by 12 gives approximately 44 ms and outlines the following limitation of the scheme validity. It is hard to imagine that in half a second the subject glanced over each of four options three times—such a speed is reminiscent of Saul Sternberg's discovery (Sternberg 1966) of the internal automatic scanning process for image recognition operating at a rate of one sample per 33–40 ms. However, this speed could not leave enough time for the slower and more complex cognitive process needed for the task in the present experiment. Most probably, on that occasion the subject glanced only once at each offer before choosing. A more detailed data examination revealed that early in the game she/he had tried only two suppliers and had remained loyal to Supplier C ever since. Therefore, a predisposition towards one option had had a powerful effect on the speed of the two fastest decisions. Although the plausibility of a twelve-look scheme may be questioned in that specific case, it still remains unproblematic from a modelling point of view, as was explained in greater technical detail in (Mengov 2014).

The entire cognitive mechanism is described by equations, most of which have already been discussed. Some of the novelties in the present study concerned the links of neurons representing the four suppliers with the dipole:

$$\frac{dy_7}{dt} = -y_7 + G[y_5]^+ + L(y_A z_{7A} + y_B z_{7B} + y_C z_{7C} + y_D z_{7D}) \tag{6.4}$$

$$\frac{dy_8}{dt} = -y_8 + G[y_6]^+ + L(y_A z_{8A} + y_B z_{8B} + y_C z_{8C} + y_D z_{8D}). \tag{6.5}$$

Equations (6.4)–(6.5) are of course adaptations of (5.15)–(5.16), debated in the previous chapter. Similarly, the emotional memories from interactions with, e.g., Supplier A, are defined as follows:

$$\frac{dz_{7A}}{dt} = y_A(-h_1 z_{7A} + [y_5]^+) \tag{6.6}$$

$$\frac{dz_{8A}}{dt} = y_A(-h_1 z_{8A} + [y_6]^+). \tag{6.7}$$

Again, Eqs. (6.6)–(6.7) are an implementation of (5.17)–(5.18).

Calibrating a separate model with the data obtained from each person constituted a computational experiment in its own right. Its first stage was to adapt the differential equations and make them amenable to computation. Secondly, a decision rule had to be constructed. Finally, the system had to be solved by means of stochastic optimization. The following sections clarify how these tasks were accomplished.

6.4.1 Solving the Differential Equations

In neuroscience, there exists of course a variety of approaches to integrate a system of differential equations. On many occasions (Grossberg and Raizada 2000; Grossberg and Seitz 2003; Grossberg and Williamson 2001), the fastest cell reactions are computed at steady state, other activity equations are solved with the Runge–Kutta–Fehlberg 4–5 method, and MTMs and LTMs are solved at a reduced time scale with Euler's method. In a model similar to the one described here, Mengov et al. (2008) implemented a version of the Runge–Kutta–Fehlberg 4–5 method with improved precision (Gammel 2004) for computing the entire system of differential equations. All of these approaches, however, are known to suffer from computational instability typical for stiff problems. The steep signal fronts are only part of the problem; another issue is the erroneous spike generation in a constant signal that happens in neural circuits containing feedback loops.

That is why I adopted a different approach (Mengov 2014) here, which sidestepped the use of numerical methods for solving differential equations altogether.

In summary, the developed algorithm introduced discrete time and computed all neuron activations at steady state while solving all long-term and medium-term memories analytically with Eqs. (5.2) and (5.4). Let us examine in more detail how this is done on the example of the neurotransmitter equation. To begin with, recall that under certain conditions, Eq. (5.2) admits a solution of the form (5.3). Those conditions involved a steady incoming signal that is changed abruptly and goes steady again. For simplicity, the change is assumed instantaneous. Another assumption is that the neurotransmitter operates at a time scale, orders of magnitude slower than the neural signals.

Under these circumstances, here I show how a recurrent solution can be derived and implemented in a computational model. The analytical solution of the neurotransmitter equation, following (5.3), is

$$z_i(t) = \frac{b_i}{b_i + c_i y_i} + c_I \exp[-t(c_i y_i + b_i)]. \tag{6.8}$$

Here, index $i = 1, 2$ designates the channels for positive and negative emotions respectively. Mind the difference between c_i, the neurobiological constant and c_I, the integration constant. Equation (6.8) says that when $t \to \infty$ the neurotransmitter quantity $z_i(t)$ asymptotically approaches $b_i/(b_i + c_i y_i)$, which depends only on the sustained signal y_i. Let at $t = t_0$ a jump in y_i occur, followed by a steady state. Then Eq. (6.8) can be rewritten as:

$$z_i(t) = \frac{b_i}{b_i + c_i y_i} + c_I \exp[-(t - t_0)(c_i y_i + b_i)]. \tag{6.9}$$

By y_i^{old} we denote the input signal before the change, and by y_i^{new} the one after it. Now let us consider a very small time interval τ, split in half by the signal jump $(0 < \tau \ll 1)$. Just before t_0 the neurotransmitter quantity in the sending neuron has habituated asymptotically and is:

$$z_i(t_0 - \tau/2) = \frac{b_i}{b_i + c_i y_i^{old}}. \tag{6.10}$$

Following Eq. (6.9), at moment $t = t_0 + \tau/2$ immediately after the change, the neurotransmitter is

$$z_i(t_0 + \tau/2) = \frac{b_i}{b_i + c_i y_i^{new}} + c_I \exp\left[-(t_0 + \tau/2 - t_0)\left(c_i y_i^{new} + b_i\right)\right]$$

$$\approx \frac{b_i}{b_i + c_i y_i^{new}} + c_I. \tag{6.11}$$

Obviously, the task is to find c_I. Because τ is infinitely small, the neurotransmitter remains virtually unchanged in it. Therefore,

$$z_i(t_0 - \tau/2) = z_i(t_0 + \tau/2). \tag{6.12}$$

We substitute both sides of Eq. (6.12) for the *rhs*-s of Eqs. (6.10)–(6.11) to obtain

$$\frac{b_i}{b_i + c_i y_i^{old}} = \frac{b_i}{b_i + c_i y_i^{new}} + c_I. \tag{6.13}$$

Equation (6.13) determines the integration constant,

$$c_I = b_i \left(\frac{1}{b_i + c_i y_i^{old}} - \frac{1}{b_i + c_i y_i^{new}} \right),$$

which can now be put into (6.9), to finally arrive at:

$$\begin{aligned}
z_i(t) &= \frac{b_i}{b_i + c_i y_i^{old}} \exp\left[-(t - t_0)\left(c_i y_i^{new} + b_i\right)\right] \\
&\quad + \frac{b_i}{b_i + c_i y_i^{new}} \left\{1 - \exp\left[-(t - t_0)\left(c_i y_i^{new} + b_i\right)\right]\right\}.
\end{aligned} \tag{6.14}$$

Equation (6.14) admits a straightforward interpretation. It expresses a gradual shift of $z_i(t)$ from equilibrium due to a steady input signal y_i^{old}, to a new equilibrium with y_i^{new}. Note that the first term includes both values of y_i. Grossberg (1984) used essentially the same formula to describe transmitter release for a new sustained input.

All neuron activities, however, are characterized by a high rate of change (see again Fig. 6.4), which makes Eq. (6.14) inapplicable for our purpose in its current form. The problematic bit is y_i^{old} in the first term—as the equation stands now it assumes $y_i^{old} = const$, which is different from the process' actual history.

Luckily, Eq. (6.14) can be adapted to account also for a fast-changing input. Let us see how. For convenience, now we introduce a discrete-time notation and reason as follows. The equation can deal with y_i^{new} for moment t, but for $t - 1$ cannot take y_i^{old} because the transmitter did not have time to habituate. However, we can introduce an equivalent $\hat{y}_i^{old}$, defined as the hypothetical signal which, had it been maintained for sufficiently long, would have put the transmitter in equilibrium, coinciding exactly with its current value $z_i(t - 1)$. An alternative history could have produced the same outcome. In that fictitious past, a "finished" habituation arrived at "equilibrium", described by

$$z_i(t - 1) = \frac{b_i}{b_i + c_i \hat{y}_i^{old}}. \tag{6.15}$$

Apparently, from Eq. (6.15) we can determine:

$$\hat{y}_i^{old} = \frac{b_i}{c_i} \frac{1 - z_i(t-1)}{z_i(t-1)}. \tag{6.16}$$

We can now substitute y_i^{old} for $\hat{y}_i^{old}$ in Eq. (6.14) and have a formula that solves the above problem. However, we are aiming at a recurrent formula and must introduce a number of other substitutions to allow for a signal jump at every time step. Thus, we set $t_0 = 0$ and $t = 1, 2, \ldots$ to account for the discrete time, and also $y_i^{new} = y_i(t)$. That is how we come to

$$z_i(t) = \frac{b_i}{b_i + c_i \frac{b_i}{c_i} \frac{1 - z_i(t-1)}{z_i(t-1)}} \exp[-(c_i y_i(t) + b_i)]$$
$$+ \frac{b_i}{b_i + c_i y_i(t)} \{1 - \exp[-(c_i y_i(t) + b_i)]\}. \tag{6.17}$$

Doing the necessary transformations in Eq. (6.17), we finally arrive at

$$z_i(t) = z_i(t-1) \exp[-c_i y_i(t) - b_i] + \frac{b_i}{b_i + c_i y_i(t)} \{1 - \exp[-c_i y_i(t) - b_i]\}. \tag{6.18}$$

Equation (6.18) constitutes a recurrent solution to the original neurotransmitter differential equation. A couple of comments are relevant here. First, this result can again be given a straightforward interpretation: It states that the new neurotransmitter value depends on the immediately preceding one, $z_i(t-1)$, as well as on the strength of the incoming signal $y_i(t)$.

Next, the computational precision of Eq. (6.18) depends critically on the size of the time step. When it is sufficiently small or in other words, when the sampling frequency is high enough to meet the demands of the Nyquist–Shannon–Kotelnikov sampling theorem, the precision is guaranteed by that theorem.

In practical terms, the question arises how exactly this time step should be determined. As already discussed, the shortest reaction time in the entire experiment was 530 ms. This had to be divided by 12 due to the three-looks-per-offer assumption, which gave 44 ms. Considering the demand of the theorem and the need for practical convenience, a time step of 10 ms (1 cs) was chosen for computing the entire system.

Further, all long-term memories—the emotional stores—had to be computed alongside the medium-term memories, and therefore a solution, similar to Eq. (6.18) was needed for them too. Reasoning exactly as above, in (Mengov 2014) the following recurrent solution to Eqs. (5.17)–(5.18) was found:

$$z_{ik}(t) = z_{ik}(t-1) \exp[-h_1 y_k(t)] + \frac{h_2}{h_1} o_l(t) \{1 - \exp[-h_1 y_k(t)]\}. \tag{6.19}$$

Here, quantity $o_l(t)$ with $l = 1, 2$ is the READ-predicted consumer emotion of satisfaction $o_1 = [y_5]^+$ or disappointment $o_2 = [y_6]^+$. Index $i = 7, 8$ indicates the particular long-term memory. Again, index $k = A, B, C, D$ designates the suppliers of omnium bonum. In simulations, it is often justified to set $h_2 = 1$.

6.4.2 Making Decisions with a Hybrid Neural Model of Intuition

Plenty was said thus far, about how the neural model was constructed. Now let us see how it can be employed to make decisions that emulate those of a human decision maker. Coming back to the omnium bonum experiment, it is clear that the current offers and the past experiences were the two major drivers behind each choice. These were not independent from each other, of course. The offers prompted immediate emotional reactions that blended with memories about supplier behaviours in the past.

One way of accounting for this situation was to construct a decision rule of three separate components (Mengov 2014). The first was the momentary neural reaction to the four omnium bonum offers. The second was the experience with each supplier accumulated in the customer's long-term memory. Finally, the third factor was the emotion, remembered after the last deal with a particular supplier regardless of how far in the past it happened. The former two factors were directly based on READ components and interactions, while the third was the result of an independent econometric study. Thus, the decision rule took this form:

$$\mathbf{D} = d_1\mathbf{F}_1 + d_2\mathbf{F}_2 + d_3\mathbf{F}_3 \tag{6.20}$$

$$K = \max(\mathbf{D}). \tag{6.21}$$

Equation (6.20) presents a weighted sum of the three factors. According to Eq. (6.21), the winning supplier K is the one who manages to obtain the highest value among four scalars—one for each supplier—in vector $\mathbf{D}$. The decision rule of the model is as simple as that. Constants d_1, d_2, d_3 add up to one and are determined in the stochastic optimization procedure alongside a set of neurobiological constants.

A watchful reader may be scandalized by this approach of introducing a rule to compute something like value/utility for each of four alternatives and then equate the largest number to the choice made. All of this is at odds with the claims about the brain not computing value by a utility function that ended Chap. 5. However, the following line of reasoning may be relevant in response to such a hypothetical objection.

First, Eqs. (6.20)–(6.21) apparently imply no utility function whatsoever. In addition, any choice between two alternatives is in the end a prevalence of one neural activity over another one. How much formal transformation to achieve some canonical representation has taken place beforehand is a question, not yet answered by science. Still smaller is the claim behind Eqs. (6.20)–(6.21), as they do not even resemble neural activity. The point in having anything like them is to offer the simplest possible decision rule around a recurrent gated dipole—a plausible neural substrate of the most primitive form of intuition, the sine qua non of the intuitive System 1.

Now let us briefly discuss each of the three factors. In more detail, they are described in (Mengov 2014).

6.4.2.1 Factor F_1: Neural Responses to Economic Options

A complicated structure as READ offers plenty of opportunities to define the neural response to the current offers. Following Mengov et al. (2008), a function based on y_7 and y_8 was developed, as these quantities integrate the two most important influences: the emotional responses $[y_5]^+$ and $[y_6]^+$, and the input from emotional memories z_{7k} and z_{8k} reflecting past experiences. In addition, the decision procedure had to accommodate the eyeballing phase with three glances at each offer. This was achieved by defining the following vectors:

$$\mathbf{y}_7^l = \left[y_7\left(t_A^l\right), y_7\left(t_B^l\right), y_7\left(t_C^l\right), y_7\left(t_D^l\right) \right]^T \tag{6.22}$$

$$\mathbf{y}_8^l = \left[y_8\left(t_A^l\right), y_8\left(t_B^l\right), y_8\left(t_C^l\right), y_8\left(t_D^l\right) \right]^T, \tag{6.23}$$

$l = 1, \ldots, n$, where l is index for the value of y_7 and y_8 during *the last of three consecutive exposures* of the dipole to each offer in each round. Here $n = 20$ is the number of rounds. In other words, $\mathbf{y}_7^l, \mathbf{y}_8^l$ are vectors containing y_7 and y_8 activations in the last four of the 12 periods of eyeballing in the lth round. The t-moments are linked to the period ends. It is a key assumption that around the last 1/3rd of the observation interval the subject already has established their opinion about the options.

There was, however, an additional complication. In simulations, it happened that due to exhausted neurotransmitter in the satisfaction channel, the feedback loops $y_7 \rightarrow y_1$ and $y_8 \rightarrow y_2$ sometimes propelled the offers to the channel for disappointment, where it was y_6 that got activated. Just like in a living human, the model's off-channel reacted to the best option (the largest amount of omnium bonum on offer) with the relatively smallest signal among the four, in effect treating it as the least unacceptable; similarly, the worst option provoked the biggest response, making it the most unacceptable. Apparently, this phenomenon is not only psychologically plausible but is also easy to understand technically, and is explained in more detail in Mengov (2014). The problem was solved by implementing simple algebraic operations, whose essence was to construct a criterion that would be channel-invariant. This was achieved in the following steps.

First, a new vector $\hat{\mathbf{y}}_8^l$ was constructed based on $\mathbf{y}_8^l$, which was "inverted" so that in the position of its largest element, in the new vector was placed the new smallest element, and vice versa:

$$\hat{\mathbf{y}}_8^l = \boldsymbol{\mu}_8^l - \mathbf{y}_8^l \qquad (6.24)$$

Here, $\boldsymbol{\mu}_8^l$ is a vector of four real numbers, all equal to the largest element, $\max(\mathbf{y}_8^l)$, of the activity vector $\mathbf{y}_8^l$. The obtained $\hat{\mathbf{y}}_8^l$ represents the 'disappointment-channel point of view' on the options. It is compatible with $\mathbf{y}_7^l$ in the sense that both have their largest, second-largest, etc. elements in the same positions. Equation (6.25) shows how both were summed up and divided by two to obtain their average, $\bar{\mathbf{y}}_{7,8}^l$, which is one form of the needed channel-invariant representation:

$$\bar{\mathbf{y}}_{7,8}^l = \left(\mathbf{y}_7^l + \hat{\mathbf{y}}_8^l\right)/2. \qquad (6.25)$$

Finally, to place all $\mathbf{F}_1$ elements between 0 and 1, vector $\bar{\mathbf{y}}_{7,8}^l$ was normed by dividing its elements by the largest among them:

$$\mathbf{F}_1 = \bar{\mathbf{y}}_{7,8}^l/\max\left(\bar{\mathbf{y}}_{7,8}^l\right) \qquad (6.26)$$

Operations (6.24)–(6.26) were certainly not the only way to neutralize the channel-alternating effect. Perhaps a more 'neuronal' and sophisticated approach could have been implemented. Yet, what was done was simple and robust—two qualities needed for the computationally intensive stochastic optimization.

6.4.2.2 Factor $\mathbf{F}_2$: Emotional Long-Term Memories and Economic Reputations

The decision function (6.20) required that all past experiences with a supplier—positive and negative—be united in a single quantity to represent its market reputation. One way of characterizing all suppliers' reputations was this:

$$\mathbf{F}_2 = 1/\left(1 + \exp\left(-\mathbf{z}_{7k}^l/\mathbf{z}_{8k}^l\right)\right), \qquad (6.27)$$

with $k = A, B, C, D$. Equation (6.27) ensures that the widely varying ratio $\mathbf{z}_{7k}^l/\mathbf{z}_{8k}^l$ of positive and negative aspects of the reputation stays constrained between 0 and 1 as can be seen in Fig. 6.4, top left plot. Alternative approaches involved other functions of the $\mathbf{z}_{7k}^l/\mathbf{z}_{8k}^l$ ratio, their logarithmic transformations, their scaled logistic transformations, LTM differences as arguments of logistic or other nonlinear functions, etc. The approach adopted here was the best compromise between simplicity and adequacy for the task.

6.4.2.3 Factor F_3: Remembered Satisfaction

Taken together, the two factors already discussed treat the emotional memory as a kind of a time-average over all interactions with a supplier. However, they have nothing to say about another type of memory about customer satisfaction: what the subject remembered from their last interaction with a supplier, no matter how far in the previous rounds this had happened. An econometric study, done by my graduate student Nickolay Georgiev and myself (Mengov and Georgiev 2013) revealed the importance of this variable. In the present model, it became a decision factor in its own right:

$$F_3 = \Psi_{last}^{t-1} \tag{6.28}$$

Here, Ψ is a vector of a participant's selectively remembered emotions of disappointment or satisfaction. The superscript in the *rhs* of Eq. (6.28) indicates that F_3 keeps track of every emotion from the first to the penultimate round $t - 1$ experienced after a supplier delivery. For example, if the game is now in its round #13, Supplier A may have been chosen for the last time in the second round when it made the participant *Very disappointed* (-3), Supplier B was last chosen in the 12th whereby the participant felt *More satisfied than disappointed* ($+1$), Supplier C in the 7th made the participant *Disappointed* (-1), and Supplier D in the 5th made the participant *As much satisfied as disappointed* (0). It is the values of the disappointment-satisfaction scale in these rounds that comprise the current content of Ψ_{last}^{t-1}. All values were rescaled from $[-4, 4]$ to $[0, 1]$.

Apparently the three factors are related, but each carries an aspect of information that is distinct from the other two. Using them together as independent variables in regression equations would of course be problematic; however, the hybrid neural model poses no such restriction. In that, it resembles the human decision maker who is helped, not hampered, by correlated factors. That is precisely why simple heuristics often perform no worse than regression techniques (Gigerenzer et al. 1999; Todd and Gigerenzer 2000).

6.4.3 Solutions by Stochastic Optimization

Now let us look at the situation from a distance. We have 18 differential equations plus a couple of algebraic ones and expect them to encompass the complexity of agent behaviour in the omnium bonum economy. Yet, they only account for the most rudimentary intuitive decision-making. Therefore, had a participant chosen in this way, the model could be expected to predict their decisions successfully. If this turned out to be true for somebody, then *that person must have chosen suppliers in the simplest affective-intuitive way without any strategizing*—otherwise the model would have failed. Therefore, anyone doing choice by intuition should be caught on the spot. People implementing mixed intuitive-strategic reasoning should be more difficult to forecast, and purely strategic thinkers should totally evade us. But let us first describe the final elements of the armoury that would allow drawing such conclusions.

Here I show how the READ-based model was made to behave—in a sense—like a human being who makes decisions. Now that the differential equations were suitably adapted, it remained to link them with the empirical data. Unfortunately, no Cauchy problem could be formulated—we have 11 equation parameters and 20 records about a participant that cannot be related analytically. A record from the ith round consists of a one-out-of-four choice, its timing $t_C^{(i)}$ indicated by a mouse click, a self-assessed emotion DS (D for disappointment, S for satisfaction), and finally its moment $t_{DS}^{(i)}$, again taken with a mouse click. We would like to have the neural model emulate the participant's answers as closely as possible.

In the spirit of the discussion in Chap. 4, a sensible way of checking the model capabilities would be to calibrate the equations with approximately half of the data and then do assessment with the other half. Naturally, the participant's records used for calibration would have to be chronologically first, while those for testing would cover the later rounds.

With no analytical means at our disposal, we had to resort to stochastic optimization. Optimal values were sought for constants $M, b_1, b_2, c_1, c_2, G, L, h_1$ in the READ equations and for constants d_1, d_2, d_3 in the decision rule. The chosen method was again simulated annealing. A suitable objective function had to be optimized to achieve the global minimum of prediction error. Consider Eq. (6.29), which was employed for the similar but simpler experiment (Mengov et al. 2008) discussed in Chap. 5:

$$J = \frac{1}{m} \sum_{i=1}^{m} I_i(t_C^{(i)}) + r_m(\boldsymbol{\Psi}(\mathbf{t}_{DS}), \mathbf{o}(\mathbf{t}_{DS})). \qquad (6.29)$$

Objective function J had to be maximized with respect to both choices and emotions. In Eq. (6.29), quantity $I_i(t_C^{(i)})$ is indicator equal to 1 if in round i the model has chosen a supplier exactly as the participant, and 0 otherwise. Here m is the number of sequential rounds taken as calibration sample. For more technical details see (Mengov et al. 2008). The second term of the equation deals with the actual self-assessed satisfactions

$$\boldsymbol{\Psi}(\mathbf{t}_{DS}) = \left[\Psi(t_{DS}^{(1)}), \dots, \Psi(t_{DS}^{(m)}) \right]^T.$$

and their model-predicted counterparts

$$\mathbf{o}(\mathbf{t}_{DS}) = \left[o_l(t_{DS}^{(1)}), \dots, o_l(t_{DS}^{(m)}) \right]^T$$

at the actual self-reported moments t_{DS}. Note that the actual emotion DS can be either positive or negative, while READ can have only positive outcomes o_1 or o_2. Therefore, to relate the empirical and computed scales one must take all o_2 values (representing disappointment) with negative signs in $\mathbf{o}(t_{DS})$.

It was necessary to put together $\mathbf{\Psi}(t_{DS})$ and $\mathbf{o}(t_{DS})$ in the objective function. A good choice was to maximize their Spearman rank correlation $r_N(\mathbf{\Psi}(t_{DS}), \mathbf{o}(t_{DS}))$, and in particular, its variant with corrections for ties in the data. Other suitable measures of association could be the Kendall rank correlation and, as long as both $\mathbf{\Psi}(t_{DS})$ and $\mathbf{o}(t_{DS})$ are at least interval-scaled, i.e., quantitative, one could also use classical correlation.

In Eq. (6.29), the first term varies within $[-1, 1]$, and the second within $[0, 1]$. As simulated annealing proceeds, J seeks to reach its maximum of 2 and thereby both terms have equal contribution.

Constructing an objective function for the omnium bonum study, however, was more demanding. With experience, it became apparent that a more direct way of dealing with nonstationarity was necessary. In particular, as the game unfolded, one could begin acting routinely and even become bored. To keep people engaged, the experimental design served them systematically changing quantities of omnium bonum, but that introduced nonstationarity and challenged the READ-based model to the limit of its capabilities.

The problem was tackled as follows (Mengov 2014). A new objective function was defined that gave advantage to the calibration rounds immediately preceding the test sample records. In particular, simulated annealing maximized this function:

$$J = \sum_{i=1}^{m_1} I_i + \gamma \sum_{i=1}^{m_2} I_{m_1+i} + R(\mathbf{\Psi}(t_{DS}), \mathbf{o}(t_{DS})). \tag{6.30}$$

Here, I_i is again a guess indicator. It is equal to 1 if in round i the model chose a supplier in the sense of Eq. (6.21) exactly as the participant, and 0 otherwise. The first summation is over the initial m_1 rounds, while the second is over the next m_2 rounds in the calibration sample. It was established that a good division comprised $m_1 = 8$ and $m_2 = 4$, which left for test sample the last eight from the entire sequence of 20.

The nonstationarity of the process was mitigated by enhancing the impact on J of the final m_2 calibration rounds, whose correct predictions were weighted more ($\gamma > 1$) than the initial m_1 rounds. Coefficient γ was chosen heuristically but not arbitrarily—it had to ensure a good balance between the two calibration subsets. Because of the crucial position of the second subset, it was decided that a combination of its all four correct choices contributing to J, together with totally incorrect choices in the first eight rounds would be valued more than another combination of all correct first eight plus only half correct among the next four. This amounted to $0 + 4\gamma > 8 + 2\gamma$, which meant that the smallest whole number to satisfy the inequality was $\gamma = 5$. This design implied that combinations of the same number of correct choices were treated differently depending on their configuration in the two calibration subsets and, say, four successes might occasionally be preferred to ten, as in the example.

Which of two solutions with identical number of guesses in the two calibration subsets should be preferred? That answer is given by the third term in Equation (6.30), the classical linear correlation $R(\mathbf{\Psi}(t_{DS}), \mathbf{o}(t_{DS}))$. Now, if two different solutions during the simulated annealing procedure yield two identical choice predictions in the 12 calibration rounds, correlation R tips the balance in favour of the one that better accounts for the participant's emotions. Thus, the objective function selects solutions fitting the calibration data with increasing precision both with regard to choices and emotions. Using the latter for tie-breaking also prevents falling in traps of local error minima during the stochastic optimization.

Following (Mengov 2013, 2014), Fig. 6.5 shows how successful that strategy was. Each person's empirical data were fit to the neural model via the optimization procedure by computing the system of equations 4,000–20,000 times. Each dot on the solid line represents a new maximum of J as computed by Eq. (6.30). The x-axis shows the number of maxima achieved during a simulated annealing run, but not the actual number of computations. Although the plots show all maxima at equal distances, in practice the game of chance may separate every two neighbours by several to several thousand computations. The black squares along the dotted lines designate errors calculated for the test sample each time the algorithm achieved a new best J for the calibration sample.

In line with Fig. 4.2 and the discussion around it, here one could expect that more calibration effort, or number of times the model is computed, would lead to higher prediction accuracy in the calibration sample, and would be accompanied by a bow-like performance in the test sample. Very often exactly that happened, as can be seen in the top four plots of Fig. 6.5. Occasionally, a new dot led to an increase (instead of reduction) of the calibration error due to the objective function discontinuity explained above. Yet, the general picture was in line with the expectations.

The top left plot shows a gradually diminishing forecasting error in calibration due to more computational effort, alongside a stubborn test sample—its prediction was successful initially only 25 % and later rose to 38 % (error falling from 0.75 to 0.63). The plot below, in contrast, tells the story of a quickly achieved test sample error of only 25 %, soon lost due to excessive fine-tuning with the calibration sample. It was a remarkable finding that somebody's choices of one among four options could be guessed with 75 % accuracy.

All the panels of Fig. 6.5 suggest that *people's predictability depended on the person being predicted*. Have a look at the bottom left plot: it presents a person whose test sample choices were estimated correctly only 25 %, which is equivalent to a random guess. The plot to the right tells the story of yet another person for whom the calibration procedure was futile, only this time the success was fifty-fifty.

In this context, the right plot in the third row deserves attention. For that person, the correct guesses from 6th to 13th improvement of J were 92 % for the calibration sample, and were 88 % in the 7th, 8th, 9th, 10th, and 13th improvement for the test sample. Which solution should be preferred? Of course, each parameter set that

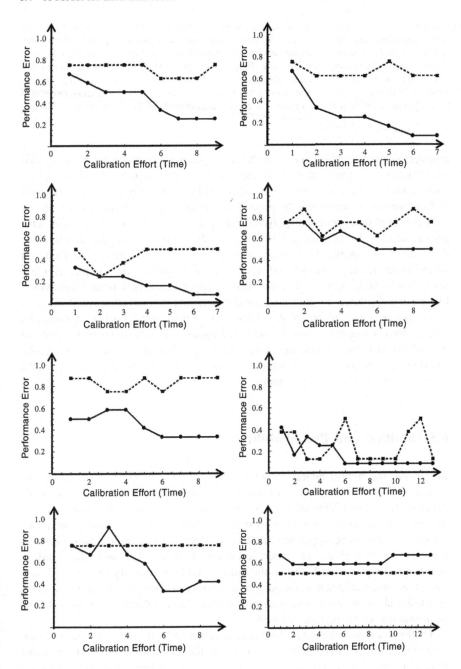

◀ **Fig. 6.5** Empirical error curves for eight anonymous individuals. The *solid line* indicates the calibration sample and the *dotted line*—the test sample. The y-axis shows the prediction error in a zero-to-one scale. A zero corresponds to correctly guessed all 12 calibration choices or all eight correctly guessed test sample choices. The abscissa gives the number of times objective function *J* achieved a new maximum during calibration. Every two neighbouring points are separated by hundreds or thousands of computations of *J* that resulted in inferior prediction success. [Reprinted from Mengov (2013), with the permission of The Bulgarian Academy of Sciences Publishing House.]

produces such marvels is good enough. The degeneracy problem already discussed does not allow a "true" set of parameter values to stand out. Should there be any game rounds beyond twenty, all of them would most likely produce very good further forecasts at least until the process nonstationarity makes them less relevant.

A couple of other people displayed thinking that was even more transparent. One person out of 131 was predicted 100 % correctly in their test sample, and 92 % in the calibration sample. Two others were predicted 88 % correctly in their test sample with respectively 83 and 100 % in their calibration sample. A close inspection of these people's data revealed no apparent anomaly in their behaviour, like staying absolutely loyal to one supplier, or acting in some other simplistic way.

Let us once again adopt a bird's eye look at what has been achieved. Essentially, here we have a couple of cases, in which somebody's choices are so predictable that we feel like pointing a magic light torch at a spot in their brain where the particular READ circuit resides. No, we have not done that yet, but the mere thought of it does not seem too fanciful any more.

6.5 Intuition and Predictability

In general, the hybrid neural model predicted with at least 75 % accuracy the choices of 8–9 % of the people in a sample of hundreds, with at least 63 % accuracy the choices of about 1/6th of all, achieved solid results with many more, and was overall useful for about two thirds of them all (Mengov 2014). Those participants, for whom it did work, apparently must have adopted a very simple approach with virtually no strategizing or expert economic thinking. Precisely that kind of decision-making is what both popular culture and contemporary science call *intuitive*. It is remarkable that a computational model consisting mostly of a handful of equations about neurons and neurotransmitters, and utilizing data unrelated to neurobiology, could perform so well.

On the other hand, the model failed to rise above 25 %, the pure guess benchmark, with about a third of all participants. Obviously, those people employed strategies far beyond its comprehension. The others in the sample—a large majority—spontaneously combined rule-based with hunch-based reasoning that lay between the two extremities. Naturally, the model predicted their choices with limited though tangible success.

As the main purpose of the study was to account for rudimentary intuitive thinking, it made sense to conduct an experiment with a slightly more complicated economic problem. To this end, my doctoral student Anton Gerunov and I developed a new design treatment with two more pieces of information on the screen. The first was "Total production of omnium bonum in the last round" and simply gave the sum of the four suppliers' potential deliveries. The second item was a forecast in percentage points about the change of omnium bonum production in the current round. Because virtually all subjects were undergraduates or graduates who had taken at least a couple of economics courses, they could easily recognize the former variable as something a lot like a country's gross domestic product (GDP) or the world production of commodities such as oil, ores, corn, etc. The latter variable was an apparent proxy for a standard economic forecast of GDP growth. While the two cues were irrelevant for the game outcome, they could trigger associations with economic theories, which in turn could lead away from amateur intuitive thinking. The point was to see how a task complication would affect choice predictability. It was not surprising that the model success dropped by about a half in the segment of highly predictable people, while the share of the impenetrable thinkers rose by a third (Mengov 2014). Complicating the game invited participants to act less intuitively. The model prediction correlated negatively ($R = -0.26$) with the amount of accumulated omnium bonum in the treatments without the two GDP proxies. It was even more so ($R = -0.36$) in the treatments with them, suggesting that the more intuition one used in the game, the less successful one was economically.

In the end, it seems justified to conclude that the Grossberg–Schmajuk dipole, augmented with a suitable decision rule, can comfortably handle intuitive choice. Tasks that are more difficult call for application of more sophisticated neural networks. However, this is a long way to travel. Neuroscience has not yet reached the level of detail that would allow us to identify a single most appropriate neural structure for cases such as the omnium bonum experiment.

The model used here was only one of three possible variants outlined in Grossberg and Schmajuk (1987). The other two are shown in Fig. 6.6 (top and centre). Moreover, Gaudiano et al. (1994) have examined a neural circuit by Raymond et al. (1992) and have concluded that it is extremely similar both functionally and structurally to READ. That network is shown in Fig. 6.6 (bottom); it could have served our purposes equally well. It might indeed be expected that in the future, research efforts like this one will rely on knowledge that is more precise, whereby establishing the right model will be a clear-cut exercise.

The computational procedure was able to capture the behaviour of some people but not of others because they reasoned differently in a systematic way. But why did they do so? Psychology has discovered that when choices are considered important, more cognitive effort is involved, while ordinary situations are tackled effortlessly and intuitively. Therefore, our subjects were heterogeneous because they perceived differently what was at stake in the game.

This last aspect of decision-making departs from the classical view of economics, positing that two agents in the same situation choose differently due to their different risk attitudes as expressed by the concavity of their utility functions. For

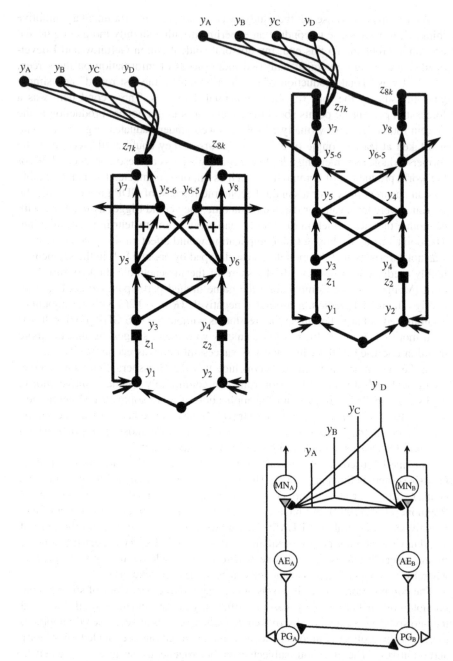

Fig. 6.6 Alternative neural networks for the experiment. *Top* and *centre* variants on the Grossberg–Schmajuk dipole. *Bottom* operant conditioning circuit by Raymond et al. (1992)

many economists the question is, How risk-averse, i.e., how afraid is the agent in the face of a risky prospect? The present study outlined a different question, which is, Is the choice important enough for strategic thinking to get involved, or would simple intuition be sufficient?

In Mengov (2014), I concluded with the observation that psychology seems to have identified at least two separate phenomena under the umbrella of "intuition". One is the most rudimentary intuition, produced by System I (Type I process) according to the mainstream dual-process theories, while the other is the expert intuition, as formulated by the dissenting fuzzy trace theory. New questions may arise out of this theoretical framework: When a choice is important enough, is the agent expert enough to address it with gist-based intuition, or must she slow down and switch to deliberative thinking. And even if she is extremely competent, perhaps the situation is too unusual and demands more than just intuition? It is hard to see any approach other than mathematical and computational neuroscience that would be capable of modelling such situations with sufficient precision.

6.6 How Humans Think They Take Decisions

To summarize a little, this study established that the simplest form of decision-making, often called intuition, is amenable to analysis and prediction by a neural model. But would the people whom that model successfully described agree to characterize their own behaviour as intuitive? A set of debriefing questions helped clarify the answer (Mengov 2014). In particular, the session started with the open question, "Please describe how you chose the suppliers of omnium bonum." The next question was, "Did you have a strategy? Yes/No." This was followed by: "If you did have a strategy, please explain when and how it was formed." The next question was, "Did you use different strategies throughout the game? Yes/No." That line of questioning was continued by, "If you did change your strategy during the game, what made you do so?" The last question of the series was, "If we gave you the opportunity to play again, would you adopt a strategy different from the one you just used?" The apparent redundancy ensured that all potentially relevant information was captured. Towards the end, a question asked about customer loyalty. Finally, the participant had to rank in a chart a number of potentially influential factors such as their immediately preceding choice, the decisions taken in the last couple of rounds, the difference between initially offered and eventually supplied omnium bonum, and some other.

A multitude of findings emerged out of the survey that would tug any researcher into deep thought and probably some amazement. First, there was the expectation that people with whom the model did well would declare acting by intuition. And indeed, comments like, "I took my decisions mostly by intuition", with variations, were often. However, essentially the same phrase appeared in the reports of many with whom the model failed miserably. There was more frustration to come.

A content analysis identified the most widely used decision strategies (Mengov 2014). A number of people were loyal to a particular supplier. Others chose only the best or the second best offer. Yet others reported avoiding the extremes, either high or low. Many were influenced by the gap between offer and delivery in the preceding round. A few admitted to have chosen randomly all the time. There were those who developed a strategy gradually and updated it throughout the game. Others coined their strategy almost immediately after the start and stuck to it ever since. Some tried to influence the suppliers' behaviour although the instruction fended off such a possibility.

Because the above is a list of clear-cut strategies, and the neural model is only good with the simplest rudimentary thinking, it is right to anticipate its complete failure with all of them. In reality, their holders populated the group of poorly predicted people just as massively as they did the group of those *well predicted.* There was no statistically significant difference between the two subsamples. In other words, what one declared and eloquently described was totally unrelated to what the READ-based model captured.

The important finding here is the inability of questions about self-assessed strategic thinking to get down to cognition, as low-level as the interaction of a handful of neurons. Intuition is there and is caught by the model, but at the same time completely evades any effort to be described in ordinary words. The set of debriefing questions was elaborate enough, yet it could not penetrate the opaque communication between the micro level of neural circuits and the macro level of human thought and language. It was refreshing that computational neuroscience managed to make a difference.

6.7 Speculative Axioms and Analytical Principles

Imagine a thought experiment, in which all decision makers belonged to the small world of the omnium bonum economy, while the real world outside the laboratory did not exist. Now, all game strategies described above are obviously rational and resting on sufficient common sense. In that simpler world, any of them would perhaps be a candidate to breed a truism, a postulate, or even an axiom. How about, "Always go for the best". Or maybe, "It is optimal to avoid the extremes". Equally sensible would be, "Judge the Supplier by its deeds, not by its words". Scholars in this fanciful realm devise theoretical concepts and build philosophical doctrines around subsets of such postulates. The more mathematically inclined develop useful axiomatic systems seeking to encompass as much empirical content as their ingenuity and analytical prowess would let them. With time, some schools of thought become more influential. The advice they give is generally reasonable, yet every now and again serious problems occur. To resolve the crises, minor theoretical revisions mend the edifice by including the troublesome new empirical findings in it in a controlled way, careful not to topple it altogether. Gradually, the participants' skills in playing the game improve, but that takes place in ever-smaller jumps. New fashionable conundrums around the omnium bonum deliveries attract the attention of the research community, to be solved by the occasional youngish scientist's marginal theoretical insight. There exists a consensus that a thorough and deep understanding leading to reliable predictions is an impossible dream.

Luckily, our world is infinitely larger and more interesting than the omnium bonum kingdom. We have developed methods beyond the means of its scholars. For a change, we could summon neuromodelling to help. From the standpoint of decision analysis, we know virtually everything there is to be known about 8–9 % of the agents and have a substantial predictive success with twice as many. In addition, it was shown (Mengov and Georgiev 2013; Mengov 2014) that over the entire test sample, the READ model beats the standard econometric tools by five percentage points. What is more important, we know exactly why we know that much, and why a lot more evades us: We have approached people's heterogeneous attitudes, person-by-person, with a minimal one-for-all neural model.

Let us once again step aside and adopt a more detached perspective on the methodological issues we must deal with. Clearly, what the fictitious omnium bonum scholars as well as many real decision scientists have in common is their reliance on a shaky theoretical basis. To be sure, studies guided by crafty postulates have given rise to tremendous achievements. Bernoulli's inception of the utility principle and Kahneman and Tversky's fundamental idea about the gains and losses as deviations from a reference point were just two among many historic milestones. Virtually any substantial psychological theory, built around insightful empirical work, has something of that kind to offer. Still more so—a recombination of such theories. In the future, we may see the utilization of ideas from social psychology— like, e.g., the ambiguity-tolerance concept—in fields demanding formal risk analysis, such as behavioural finance, human-machine interactions, etc. Less popular ideas may be just as fruitful. Management science has made the most of the overconfidence concept, but its exact opposite—the lack of self-confidence—is no less important. For example, Zinovieva (1989) studied how people reason when dealing with complicated decision tasks. It turned out that confident people rely a lot on their own interim conclusions to reach a decision efficiently. Those lacking confidence, in contrast, tend to assign equal value to all relevant alternatives, thus obstructing and even frustrating the decision-making process. To them, very different prospects look equally attractive.

Any such interesting discovery will constitute a reason to rethink and enrich ad infinitum the fundamental postulates and the various axiomatic systems of decision science. However, that approach would never get us out of the predicament of the fictitious scholars above. Its main fault is that it seeks to theorize with intermediary and secondary concepts while pretending that any subset of those is somehow primary and fundamental. This enterprise is not unlike the ancient Greeks' attempt to explain the entire world with the five Platonic solids. Instead, we would better heed to the example of the astronomer Kepler, who tried to relate them to the solar system, but abandoned them to discover new laws of planetary motion.

This discussion leads to the question, what could constitute a "Newtonian mechanics" for the brain: A small set of principles—and their analytical formulations—that are universal enough. Looking back at Grossberg's two differential equations of neural interaction and the Hodgkin–Huxley equation of neural activity, together they appear as a very strong candidate. It is unlikely that any psychological or behavioural finding could shake them—they are simply too low-level. They can be extended, combined and recombined, to model great many cognitive phenomena, as has been shown many times already.

In this chapter, I demonstrated how a model, built from these three equations, penetrated the thinking of a small minority of people—those who resort to quick intuition. Future modelling and experimenting would bring further success. It is more important, however, that one can hardly see how a conceptual obstacle might stop this effort. Surely, these equations—just like the real Newtonian mechanics in physics— will not be enough to explain everything about cognition. No doubt, more such principles would be needed. A currently missing postulate can perhaps be formulated about spiking neurons or about something else. All the same, contemporary decision science has at its disposal more means than it seems to recognize at present.

References

Gigerenzer, G., Todd, P., & the ABC Research Group. (1999). *Simple heuristics that make us smart*. Oxford: Oxford University Press.

Grossberg, S. (1984). Some psychophysiological and pharmacological correlates of a developmental, cognitive, and motivational theory. In R. Karrer, J. Cohen, & P. Tueting (Eds.), *Brain and information: Event related potentials* (pp. 58–142). New York: New York Academy of Sciences.

Grossberg, S., & Raizada, R. (2000). Contrast-sensitive perceptual grouping and object-based attention in the laminar circuits of primary visual cortex. *Vision Research, 40*, 1413–1432.

Grossberg, S., & Schmajuk, N. (1987). Neural dynamics of attentionally-modulated pavlovian conditioning: Conditioned reinforcement, inhibition, and opponent processing. *Psychobiology, 15*(3), 195–240.

Grossberg, S., & Seitz, A. (2003). Laminar development of receptive fields, maps, and columns in visual cortex: The coordinating role of the subplate. *Cerebral Cortex, 13*, 852–863.

Grossberg, S., & Williamson, J. R. (2001). A neural model of how horizontal and interlaminar connections of visual cortex develop into adult circuits that carry out perceptual groupings and learning. *Cerebral Cortex, 11*, 37–58.

Leven, S., & Levine, D. (1996). Multiattribute decision making in context: A dynamic neural network methodology. *Cognitive Science, 20*, 271–299.

Levine, D. (2006). Neural modelling of the dual motive theory in economics. *The Journal of Socio-Economics, 35*, 613–625.

Levine, D. (2009). Brain pathways for cognitive-emotional decision making in the human animal. *Neural Networks, 22*, 286–293.

Levine, D. (2012). Neural dynamics of affect, gist, probability, and choice. *Cognitive Systems Research, 15, 16*, 57–72.

Mengov, G. (2013). Stochastic calibration of a neurocomputational model for economic decision-making. *Proceedings of the Bulgarian Academy of Sciences (Comptes Rendus), 66*(5), 739–748.

Mengov, G. (2014). Person-by-person prediction of intuitive economic choice. *Neural Networks, 60*, 232–245.

Mengov, G., & Georgiev, N. (2013). Econometric vs. ARTMAP prediction of economic choice. *Proceedings of the Bulgarian Academy of Sciences (Comptes Rendus), 66*(3), 415–422.

Mengov, G., Egbert, H., Pulov, S., & Georgiev, K. (2008). Affective balances in experimental consumer choices. *Neural Networks, 21*(9), 1213–1219.

Raymond, J., Baxter, D., Buonomano, D., & Byrne, J. (1992). A learning rule based on empirically derived activity-dependent neuromodulation supports operant conditioning in a small network. *Neural Networks, 5,* 789–803.

Rottenstreich, Y., & Hsee, C. (2001). Money, kisses, and electric shocks: On the affective psychology of risk. *Psychological Science, 12,* 185–190.

Sternberg, S. (1966). High-speed scanning in human memory. *Science, 153*(3736), 652–654.

Todd, P., & Gigerenzer, G. (2000). Précis of simple heuristics that make us smart. *Behavioural and Brain Sciences, 23,* 727–780.

Zinovieva, I. L. (1989). *The impact of confidence on the organization of cognitive activity.* Doctoral Dissertation, Department of General Psychology, Moscow State University (In Russian).

Zizzo, D. J. (2013). Claims and confounds in economic experiments. *Journal of Economic Behavior & Organization, 93,* 186–195.

Chapter 7
Adaptive Resonances Across Scales

7.1 Social and Neural Networks

As Stephen Grossberg developed his famous equations of cooperative-competitive neural interactions, he briefly examined their applicability for characterizing economic systems. He showed that some of his models described equally well neurons competing locally while exhibiting globally coordinated behaviour, and production companies driven by Adam Smith's "invisible hand" in a class of stable competitive markets (Grossberg 1980a, b, 1988). Interestingly, at macroscopic level some systems seemed cooperative while in reality they were competitive. Whether the competing components could ultimately begin to cooperate to establish structures that are more complex remained an open question. At the time, that line of research was pursued no further, but in the age of virtual social networks and big data analysis it may gain renewed importance.

Considering that the invisible hand is a mix of market signals related to prices, perceived demand, customer opinions, company reputations etc., all of them enhanced by the speed of modern communications, one can view today's economy to a large extent as a virtual social network. The link between the fields of social networks and neural networks was perhaps best summarized by Bruno Apolloni'sremark that, "The social network is a fractal extension of our brain networks" (Apolloni 2013). This is a modern variation of the old idea about monadology by Leibniz (1714), who suggested that each living creature, plant or animal, contains in itself a multitude of its own micro replicas.

This book ends with a discussion about an analogy between the operations in the adaptive resonance theory (ART) neural network (See Box 7.1) and some of the essential procedures in leader-electing social organizations. The possible foundations for such a common mechanism are examined, as well as the implications for the social sciences of a knowledge transfer from mathematical and computational neuroscience.

© Springer-Verlag Berlin Heidelberg 2015
G. Mengov, *Decision Science: A Human-Oriented Perspective*,
Intelligent Systems Reference Library 89, DOI 10.1007/978-3-662-47122-7_7

Box 7.1. Adaptive Resonance Theory

Scholars in ancient times discovered that people distort reality when perceiving it. In the 1st century C.E., the Greek philosopher Epictetus noticed that humans are affected not by events happening around them, but by their own attitudes to those events. Indian gurus made a similar observation by saying that, "You cannot see more than what you are". As society developed, that idea began to receive scientific garments. Johannes Mueller, a 19th century physiologist and mentor of Herman von Helmholtz, proclaimed in 1826 that we do not comprehend what we see directly, but only absorb our own neural responses to external stimuli. Helmholtz (1866, 1896) combined theoretical and experimental methods to develop his theory of unconscious inference. It stated that people learn new knowledge only after their senses modify all incoming information under the guiding influence of previous knowledge. In other words, we perceive and learn what we expect to perceive, based on previous experience and education.

Scholars from the humanities and social sciences have often come across the same insight. It is present in the works of prominent figures such as art historians Gombrich (1972, 1989) and Bell (2007) and science historian Kuhn (1962) . It is also popular in the folklore of various professions. Earlier in the book, I quoted human resource managers who claimed that, "Reality is not the facts, it is the interpretation of facts". Financier George Soros observed a similar phenomenon in the capital markets: the agents' beliefs affect the fundamentals behind share prices, whereby reality is driven away from those beliefs. The latter gradually become inadequate and need updating (Soros 1988, 1995).

This mindset is summarized scientifically by adaptive resonance theory (ART), initially introduced by Grossberg (1976a, b, 1980a, b, 1982), and further developed in cooperation with Gail Carpenter (Carpenter and Grossberg 1987a, b, 1990) and others (Carpenter et al. 1991a, b, 1992, 1998, etc.). It is built from the same three differential equations or their algebraic approximations. According to this theory, all knowledge is stored in connections among neurons in the brain whereby, to a first approximation, three neural layers are instrumental. They exchange signals in two directions: a "bottom-up" stream comes in from the senses and provokes a "top-down" response of associations, based on previous knowledge. Both streams are compared and matched to produce "impressions", which, if found adequate in a certain mathematical sense, are eventually memorized. These interactions are shown in the Figure.

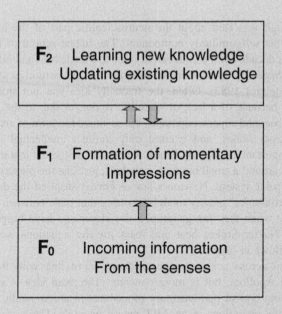

Grossberg introduced a model clarifying how one manages to learn new knowledge without destroying the existing. It posits that the brain is "plastic" as it is able to accommodate change, and at the same time "stable" as it retains what has been learned before. This is the solution to the famous "stability-plasticity dilemma". It is the object of adaptive resonance theory (ART) and its central element—the ART neural network. The term "adaptive resonance" denotes information processing and is analogous to the physical resonance in mechanical and electrical systems. It is *information* that "resonates", as multidimensional signals are exchanged between layers F_1 and F_2 in the Figure. Responding to an incoming image, the neural network instantly scans its memories to find a sufficiently close match. If one exists, all related neurons are activated to exchange signals with the impressions layer. This process is called adaptive resonance. Until it lasts, knowledge update takes place. The interaction is local as it affectsa limited number of synaptic connections. If the old memories fail to offer an adequate match, a new set of neurons assimilates the incoming signals and patterns, whereby the network enters again a resonant state.

7.2 A Fractal-Type Analogy

Now that enough was said about the neuroscientific part of the alliance, let us discuss briefly its self-similarity component. The fractal school of thought flourished in the last decades of the 20th century due to Mandelbrot and others, and gave fruits in the shape of research methodologies for the natural sciences and life sciences (Mandelbrot 1983). Often, the fractality idea was not straightforwardly utilized, either because of a lack of scientific rigour, or due to the huge distance between the subset and superset domains, but assumed the form of analogy between phenomena across scales, and exerted only indirect intellectual influence. For example, an important analogy by Ernest Rutherford suggested that the electrons in the atom circle around a small but heavy nucleus, just like the planets move around the sun in the solar system. Newton's law of gravity inspired the development of international economics' gravity models, positing that trade between two countries is more intense when they are geographically closer and have bigger economies. Similarly, the Navier–Stokes heat and mass transfer equations were adapted to model capital flows in finance.

The analogy across scales, suggested here, is in line with the insights of Grossberg and Apolloni, but is more concrete. The main idea is summarized in Fig. 7.1, showing typical parliamentary procedures in a democratic establishment that resemble the operations in an ART neural network. There are many similar details between the two sequences of events.

The left column in Fig. 7.1 describes political events characterizing parliamentary democracies. The civilized manner in which these governments, their leaders, and members, replace each other tends to disguise the intense and often fierce power struggle behind the scene. The example here is modelled after the typical contemporary European country, yet it could be easily reshaped in line with the procedures in North America or in the ancient democracies of the Greek cities and Rome. With some adaptation, the chart would comfortably fit the cardinals' conclave electing a new Pope. Similar in principle are the ways in which corporations and all kinds of institutions, large and small, replace their chief executive officers, presidents, commanders-in-chief, deans, editors-in-chief etc. Moreover, all totalitarian dictatorships also have their own mechanisms for power transfer that could be accommodated by variations of the left column in Fig. 7.1. Hardly different, though simpler, is the power handover in the animal world, where each species has developed its own rituals—generally brutal—to determine the next leader of the herd.

The right column in Fig. 7.1 describes the operations in the adaptive resonance theory (ART) neural network as they happen in time. Today we know that adaptive resonances are widespread in the brain (Grossberg 2013) and take part in many cognitive processes. Recalling Stephen Jay Gould's vision that the human brain was not build for a restricted purpose, but as it evolved for hunting, social cohesion and other functions, it transcended the adaptive boundaries of its original purpose (Gould 1981), it seems plausible that the mechanism of adaptive resonance could have extended to interpersonal relations. That is how and why individuals in a

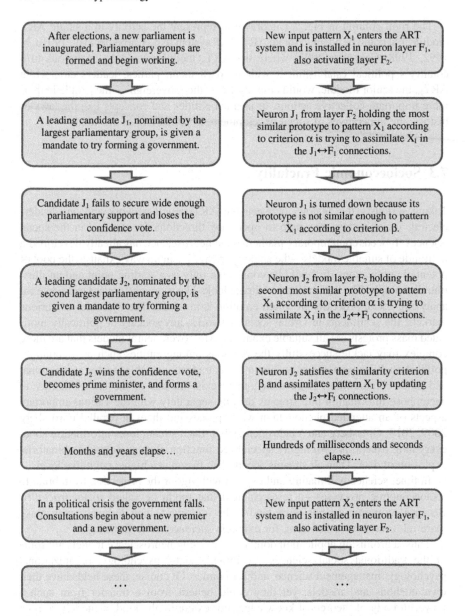

Fig. 7.1 An analogy between parliamentary procedures and neural network interactions. The way in which political leaders are elected resembles the way neurons in the F_2 layer of an ART neural network get activated to accommodate new knowledge

social network, such as the political system or any other socioeconomic system with a hierarchy, may resemble neurons in an ART neural network.

The right column in Fig. 7.1 is drawn to represent the activity in a single ART module, but could be adapted for an ARTMAP system (e.g., Carpenter et al. 1991a, b, 1992, 1998; Carpenter 2003) where the ART$_a$ module would be analogous to a country's political establishment with its parliament, parties, leaders etc., while ART$_b$, the senior module, would embody either the "sovereign"—the people in their more important collective actions, or abstract entities and concepts like the laws of history, the imperatives of social development, or something of this kind.

7.3 Socioeconomic Fractality

Taking seriously the analogy between the ART neural network and any leader-electing social system may help to open new directions for research in the social sciences. Indeed, contemporary mathematical neuroscience has already developed a multitude of suitable models—the analogy in Fig. 7.1 in no way exhausts the pool of potential applications. It should be considered only as a starting point and an illustrative example. The recurrent gated dipole looks like a straightforward candidate for analyzing social processes in which emotion is involved. The dynamics of various markets, the sentiments in virtual social networks, any socially or politically motivated mass protests are all suitable examples. Moreover, neural models that are more complex may tackle successfully the relations among entities such as a country's political establishment, industry, labour force, trade unions, third sector, professional and other communities. Even today, there exist models of the brain's ability to decompose information into streams dealing separately with the various important aspects of an attended phenomenon while processing them in parallel (Grossberg 2009, 2013; Grossberg and Vladushich 2010). Each stream loses information about everything except its own highly specialized function, thus avoiding combinatorial explosion of data. Then all streams accomplish fusion at a higher cognitive level.

In time, science will mature and explain with rigour the projection from brain to socioeconomic processes—a research field that might be called *social fractality*. Because economic aspects often tend to be important, their influence may make more relevant some other term, for example *socioeconomic* fractality.

Thus, a promising application domain for existing neuroscience models are some of the traditional social sciences—sociology, social psychology, organizational psychology, management science, and economics. Of course, these fields have their own methods and models, yet they could benefit from a transfer from such a powerful body of theoretical knowledge. For example, the kind of models coming from the Grossberg School could be used to forecast the evolution of important trends and events in socioeconomic systems, previously unpredictable. And if prediction may seem too ambitious, another important goal is to generate new philosophical explanations of poorly understood social phenomena from the past or the present. For instance, in the 1990s, the peoples in Eastern Europe embraced democracy with its key attributes such as free elections—now taking place frequently—and parliamentary procedures exactly as in Fig. 7.1's left column. However, they were soon disappointed by the hardships of economic reforms leading to pain and suffering. Putting that social development side by side with neural network operations shows it in a new perspective: It now looks like those societies had undertaken a learning process on a grand scale and were taking only the first steps in a journey of historic proportions.

A different but related new field for applying the models of mathematical and computational neuroscience may become the study of virtual social networks. Research in that area has gained momentum not least due to the invasion of Facebook, Twitter, LinkedIn and other internet platforms that have already become household names. In response, over the years many scientific journals devoted special issues to the subject, while leading international publishers even launched new dedicated journals. However, their content has remained mostly empirically oriented, a fact suggesting that a deep theoretical grasp is hard to achieve.

The future may bring about new alliances between neuromodelling and web data analyses. In fact, such examples already exist. A study by Sakata and Yamamori (2007) revealed a topological similarity between the brain and some social networks. It was based on positive and negative influences among the participating units—i.e., neurons and people respectively. That effort quantified some of the realistic boundaries of the analogy between the two domains.

There are essential pragmatic aspects of the suggested knowledge transfer. It is generally believed that the collective mind is less sophisticated than the single mind, not least because the brain has orders of magnitude more elements (neurons) and connections than any social network. Therefore, the advocated foray into socioeconomic systems should begin with some of the simpler neural models such as gated dipoles, ART and ARTMAP networks. A major difficulty would be to identify prospective concepts and variables from the social sciences that could be mapped onto suitable components of the neuroscience models. This nontrivial task may take quite long. Yet, as was discussed already in the previous chapters, some small steps have already been taken by a number of researchers.

A hypothetical neuroscience-inspired social science model could look like the creation of some branches of contemporary theoretical physics: a mathematical structure nicely fitting key elements from the general picture, yet containing quantities about which little or nothing is known. However, phenomenological and semi-empirical models are the natural early companions of each pioneering effort. All the same, if the hypothesized link between neural systems operating in the millisecond-to-second range and socioeconomic systems evolving over months and years becomes the object of intense research, it may open immense opportunities for a new kind of social science.

References

Apolloni, B. (2013). Toward a cooperative brain: continuing the work with John Taylor. doi:10.1109/IJCNN.2013.6706715.

Bell, J. (2007). *Mirror of the world: A new history of art*. New York: Thames and Hudson.

Carpenter, G.A. (2003). Default ARTMAP. *Proceedings of the International Joint Conference on Neural Networks (IJCNN'03)*, (pp. 1396–1401). Portland, Oregon.

Carpenter, G. A., & Grossberg, S. (1987a). A massively parallel architecture for a self-organizing neural pattern recognition machine. *Computer Vision, Graphics, and Image Processing, 37,* 54–115.

Carpenter, G. A., & Grossberg, S. (1987b). ART-2: Self-organization of stable category recognition codes for analog input patterns. *Applied Optics, 26,* 4919–4930.

Carpenter, G. A., & Grossberg, S. (1990). ART-3: Hierarchical search using chemical transmitters in self-organizing pattern recognition architectures. *Neural Networks, 3,* 129–152.

Carpenter, G. A., Grossberg, S., & Rosen, D. B. (1991a). Fuzzy ART: Fast stable learning and categorization of analog patterns by an adaptive resonance system. *Neural Networks, 4,* 759–771.

Carpenter, G. A., Grossberg, S., & Reynolds, J. H. (1991b). ARTMAP: Supervised real-time learning and classification of nonstationary data by a self-organizing neural network. *Neural Networks, 4,* 565–588.

Carpenter, G. A., Grossberg, S., Markuzon, N., Reynolds, J. H., & Rozen, D. B. (1992). Fuzzy ARTMAP: A neural network architecture for incremental supervised learning of analog multidimensional maps. *IEEE Transactions on Neural Networks, 3,* 698–713.

Carpenter, G. A., Milenova, B. L., & Noeske, B. W. (1998). Distributed ARTMAP: A neural network for fast supervised learning. *Neural Networks, 11,* 793–813.

Gould, S. J. (1981). Hyena myths and realities. *Natural History, 90*(2), 16.

Gombrich, E. H. (1972, 1989). *The story of art.* Oxford: Phaidon Press.

Grossberg, S. (1976a). adaptive pattern classification and universal recoding: I. Parallel development and coding of neural feature detectors. *Biological Cybernetics, 23,* 121–134.

Grossberg, S. (1976b). Adaptive pattern classification and universal recoding: II. Feedback, expectation, olfaction, illusions. *Biological Cybernetics, 23,* 187–202.

Grossberg, S. (1980a). Biological competition: Decision rules, pattern formation, and oscillations. *Proceedings of the National Academy of Sciences, 77*(4), 2338–2342.

Grossberg, S. (1980b). How does a brain build a cognitive code? *Psychological Review, 87,* 1–51.

Grossberg, S. (1982). *Studies of mind and brain: Neural principles of learning, perception, development, cognition, and motor control.* Boston, MA: Reidel Publishing Co.

Grossberg, S. (1988). Nonlinear neural networks: Principles, mechanisms, and architectures. *Neural Networks, 1,* 17–61.

Grossberg, S. (2009) Cortical and subcortical predictive dynamics and learning during perception, cognition, emotion, and action. Philosophical Transactions of the Royal Society of London, special issue *Predictions in the brain: Using our past to generate a future, 364,* 1223–1234.

Grossberg, S. (2013). Adaptive resonance theory: How a brain learns to consciously attend, learn, and recognize a changing world. *Neural Networks, 37,* 1–47.

Grossberg, S., & Vladushich, T. (2010). How do children learn to follow gaze, share joint attention, imitate their teachers, and use tools during social interactions? *Neural Networks, 23,* 940–965.

Helmholtz, H. von (1866, 1896) *Handbuch der Physiologischen Optik.* Hamburg und Leipzig: Voss.

Kuhn, T. (1962, 1970) *The structure of scientific revolutions.* Chicago: The University of Chicago Press.

Leibniz, G.W. (1714) The Monadology. Translated by George MacDonald Ross, 1999 (quotation from §67).

Mandelbrot, B. (1983). *The fractal geometry of nature.* San Francisco: W.H. Freeman.

Sakata, S., & Yamamori, T. (2007). Topological relationships between brain and social networks. *Neural Networks, 20,* 12–21.

Soros, G. (1988). *The alchemy of finance.* New York: Simon & Schuster.

Soros, G. (1995). *Soros on Soros: Staying ahead of the curve.* New York: Wiley.

Index

© Springer-Verlag Berlin Heidelberg 2015
G. Mengov, *Decision Science: A Human-Oriented Perspective*,
Intelligent Systems Reference Library 89, DOI 10.1007/978-3-662-47122-7

Printed in the United States
By Bookmasters